T0419129

Milestones in Drug Therapy

For further volumes:
http.//www.springer.com/series/4991

Irene M. Ghobrial • Paul G. Richardson •
Kenneth C. Anderson
Editors

Bortezomib in the Treatment of Multiple Myeloma

Volume Editors
Irene M. Ghobrial
Dana-Farber Cancer Institute
Boston, MA 02115, USA
irene_ghobrial@dfci.harvard.edu

Paul G. Richardson
Dana-Farber Cancer Institute
Boston, MA 02115, USA
Paul_Richardson@dfci.harvard.edu

Kenneth C. Anderson
Dana-Farber Cancer Institute
Boston, MA 02115, USA
Kenneth_Anderson@dfci.harvard.edu

Series Editors
Prof. Dr. Michael J. Parnham
Director of Science & Technology
MediMlijeko d.o.o.
Pozarinje 7
HR-10000 Zagreb
Croatia

Prof. Dr. Jacques Bruinvels
Sweelincklaan 75
NL-3723 JC Bilthoven
The Netherlands

ISBN 978-3-7643-8947-5 e-ISBN 978-3-7643-8948-2
DOI 10.1007/978-3-7643-8948-2

Library of Congress Control Number: 2010938799

Cover illustration: The ubiquitin-proteasome pathway; see Fig. 1 in the chapter of Alfred L. Goldberg "Bortezomib's Scientific Origins and Its Tortuous Path to the Clinic"

Cover design: deblik, Berlin

Printed on acid-free paper

Springer Basel AG is part of Springer Science + Business Media (www.springer.com)

Preface

The proteasome is a highly conserved multicatalytic protease that is responsible for cellular protein turnover, and several therapeutic agents have been developed that specifically target the proteasome. This truly targeted therapy has significantly altered the management of patients with Multiple Myeloma and improved survival. The prototype and first-in-class of clinical usable proteasome inhibitors is the therapeutic agent bortezomib, a boronate peptide, which reversibly inhibits the 20s subunit and has shown efficacy in Multiple Myeloma and other hematological malignancies, including Mantle cell lymphoma, primary systemic amyloidosis, and Waldenstrom Macroglobulinemia. Other proteasome inhibitors that target different subunits of the proteasome have been developed, some of which have distinctly different toxicity profiles compared to bortezomib, in part due to different chemical scaffolds as well as the ability to irreversible bind to the proteasome.

In this book, the history of the discovery and mechanism of proteasome inhibitors is described. The preclinical activity of bortezomib in Multiple Myeloma is also examined, not only on the tumor clone itself, but also on the microenvironment including its activity on osteoclasts and osteoblasts. We then focus on the therapeutic application of bortezomib in patients with newly diagnosed Multiple Myeloma as well as relapsed Multiple Myeloma. Toxicities observed with bortezomib and how to manage them are outlined, specifically focusing on neuropathy. Finally, we describe the activity of bortezomib in other hematological malignancies including Mantle cell lymphoma, Waldenstrom's Macroglobulinemia, and primary systemic amyloidosis. The book concludes by giving the readers a view of future applications of proteasome inhibitors and new generations of proteasome inhibitors that are currently being tested in preclinical or early clinical trials in cancer therapy.

As editors, we feel privileged to have participated in this exciting journey of the proteasome discovery and the application of targeted therapy into the clinical practice. We believe that proteasome inhibition has been one of the most remarkable success stories of oncologic drug discovery from the basic understanding of the proteasome and its function to the remarkable activity observed in our patients, culminating in the survival benefit that they enjoy with these agents. We hope that this book can provide useful information for the readers who want to know more about the application of proteasome discovery from the lab to the clinic.

Boston

Irene M. Ghobrial
Paul G. Richardson
Kenneth C. Anderson

Contents

Bortezomib's Scientific Origins and Its Tortuous Path to the Clinic

Alfred L. Goldberg

Abstract The development of bortezomib for the treatment of multiple myeloma was made possible by multiple major advances in our understanding of intracellular protein breakdown. The primary route for degradation of intracellular proteins is the ubiquitin–proteasome pathway, in which protein substrates are linked to ubiquitin chains, which marks them for degradation by the large proteolytic complex, the 26S proteasome. It utilizes ATP to unfold protein substrates and inject them into the 20S proteasome where proteins are digested to small peptides. Its active sites cleave proteins by an unusual mechanism that allows their selective inhibition by bortezomib. This molecule was generated by a small biotechnology company whose initial goal was to synthesize proteasome inhibitors to reduce the excessive proteolysis that causes muscle atrophy and cachexia. However, the availability of proteasome inhibitors suggested other therapeutic possibilities; for example, its role in the activation of the NF-κB suggested anti-inflammatory and antineoplastic actions and led to clinical trials against various cancers. Indications of the special sensitivity of multiple myeloma were observed in Phase I trials, and bortezomib was initially approved after only Phase II trials. In myeloma cells, NF-κB is particularly important for cell growth, but bortezomib has multiple antineoplastic actions, especially its ability to inhibit the selective degradation of misfolded proteins, and myeloma cells are continually generating such abnormal immunoglobins. Although its development was unusually rapid, these efforts encountered many obstacles, and several times were almost terminated. Beyond their major clinical impact, proteasome inhibitors are extremely useful research tools and have enabled enormous advances in our understanding of cell regulation, immune mechanisms, and disease.

A.L. Goldberg
Department of Cell Biology, Harvard Medical School, Boston, MA 02115, USA
e-mail: alfred_goldberg@hms.harvard.edu

I.M. Ghobrial et al. (eds.), *Bortezomib in the Treatment of Multiple Myeloma*,
Milestones in Drug Therapy,
DOI 10.1007/978-3-7643-8948-2_1,

1 Introduction

Probably the greatest reward that a life in biochemical research can provide is the satisfaction that comes from seeing one's work lead to a greater understanding of the functioning of living organisms and even, on occasion, to improvements in medical care. Consequently, those of us whose research has focused on the mechanisms of intracellular protein breakdown and the functions of the proteasome view with particular satisfaction and some parental pride the exciting success of proteasome inhibitors in the treatment of hematological cancers. Also highly gratifying has been the enormous utility that proteasome inhibitors have had as research tools that have greatly expanded our knowledge of the importance of the ubiquitin–proteasome pathway in cell regulation, immune surveillance, and human disease (Fig. 1).

The development of proteasome inhibitors culminating in the synthesis of bortezomib (Velcade™; originally known as PS-341) and its introduction for the treatment of cancers has had a curious and unpredictable history that is very much linked to the multiple strands of my own research career. This chapter reviews the

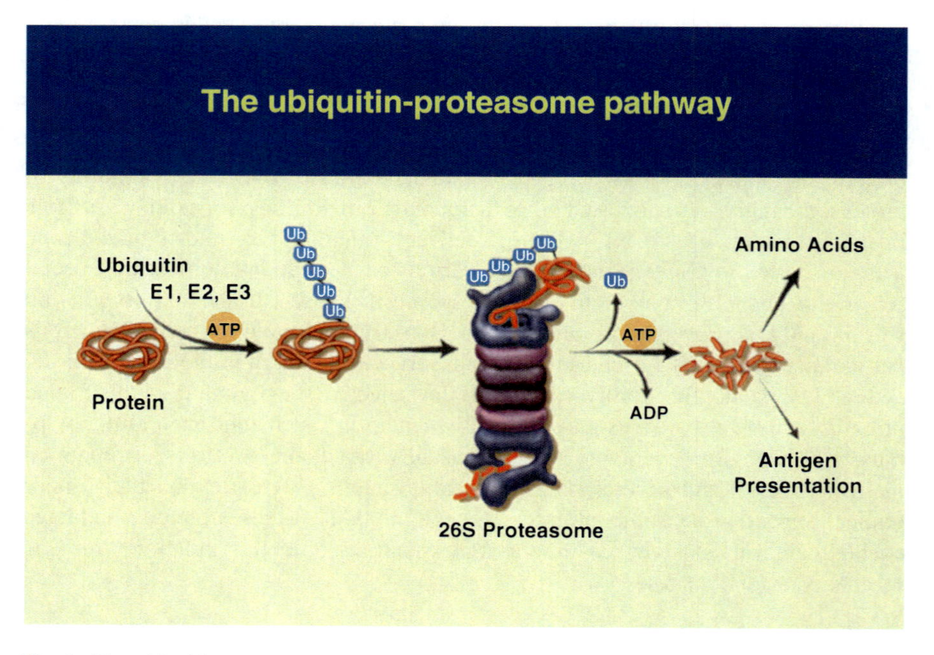

Fig. 1 The ubiquitin–proteasome pathway. This scheme illustrates the three main steps in the pathway: (1) ubiquitination, which actually involves hundreds of different ubiquitination enzymes, (2) proteasomal degradation, and (3) hydrolysis of the peptides generated. Not shown are the many deubiquitinating enzymes functioning in mammalian cells and additional factors (e.g., p97/VCP complex or "shuttle factors") that facilitate the delivery of ubiquitinated proteins to the 26S complex

scientific findings that led to the development of bortezomib. Although this agent has had a dramatic impact on the treatment of multiple myeloma, we were not aiming to find new cancer therapies when we initiated this research and organized a biotechnology company whose primary goal was to generate inhibitors of the ubiquitin–proteasome pathway. However, advances in biochemistry and development of new drugs often do not follow predictable paths, and new insights in this area, as in others, have often come about by unexpected routes. This chapter, in addition to explaining the scientific background and history for the preclinical development of proteasome inhibitors, attempts to summarize our present understanding of the proteasome's function that has emerged from 25 years of biochemical investigations, including studies during the past 18 years using proteasome inhibitors.

Our original decision in 1992 to found a biotech company and to undertake an effort to synthesize proteasome inhibitors for possible therapeutic applications was not made in the hope of advancing the treatment of cancer. Instead, it was based upon insights emerging from more arcane studies I had initiated over 40 years ago as a medical student to clarify the mechanisms of muscle atrophy, such as that occurring with denervation or disuse. This early work demonstrated that loss of cell protein following denervation was caused not by a simple reduction in muscle protein synthesis (as we had expected), but rather by an overall acceleration of the rate of protein breakdown [1]. This finding was the first evidence that overall rates of protein breakdown in mammalian cells are regulated and can be of major importance in human disease. More recent work then indicated that a similar acceleration of protein breakdown causes the muscle wasting seen in fasting and in many major systemic diseases, including cancer cachexia, sepsis, renal failure, cardiac failure, and AIDS [2, 3]. Based upon this insight, it seemed possible to develop an agent that might inhibit this process and be useful in treating this wide array of catabolic conditions.

When the initial evidence became apparent that overall protein breakdown rates were regulated, virtually nothing was known about the pathways for protein catabolism in cells [4, 5]. Therefore, in starting my own laboratory in 1969, I decided to focus my research on elucidating this degradative pathway through biochemical studies, not in muscle, but rather in simpler cells, initially *Escherichia coli* (because of the opportunity to use genetic approaches) [5, 6] and then mammalian reticulocytes [7]. In the 1970s and 1980s, these studies reviewed below led to our discovery that in addition to the lysosome, eukaryotic cells contained a soluble ATP-dependent proteolytic system [8, 9]. Analysis of this degradative system by Avram Hershko, Aaron Ciechanover, and Ernie Rose led to the discovery of the involvement of ubiquitin in this pathway [8, 10–13] for which they received the Nobel Prize in chemistry in 2005 and also to the subsequent discovery by Rechsteiner's laboratory and our own of the 26S proteasome (see below) [14–17].

Although seemingly unrelated to muscle wasting or cancer therapy, these very basic studies have eventually had enormous influence on both areas. In the early 1990s, these biochemical discoveries and parallel findings about muscle protein degradation suggested to us that selective pharmacologic inhibition of the ubiquitin–proteasome pathway might be feasible and could provide a rational approach toward developing agents to combat muscle wasting and cachexia. In pursuing that objective, many important scientific insights have been obtained, and most excitingly, bortezomib has emerged as a very valuable new treatment for hematologic malignancies.

2 Discovery and Key Features of the Ubiquitin–Proteasome Pathway

A major motivation for our undertaking to study protein degradation in bacteria and reticulocytes was that these cells lack lysosomes, which were then believed to be the site of protein breakdown in cells. However, I had become convinced that these organelles could not account for the extreme selectivity and regulation of intracellular proteolysis [5]. Our present understanding of this process emerged from a series of developments, first through key in vivo findings and then in vitro studies of their biochemical basis. Of particular importance was my finding that bacteria, which were long believed not to degrade their own proteins, could not only activate intracellular proteolysis when starved for amino acids or aminoacyl tRNA [6], but that even during exponential growth, these cells rapidly degraded proteins with abnormal structures, such as those that may arise by mutation or postsynthetic damage [7]. Reticulocytes were also shown to carry out the degradation of abnormal proteins, which further suggested to us the existence of a non-lysosomal system for selective protein breakdown in eukaryotic cells [5].

We showed that a key feature of this process was that it required ATP [5, 7, 9], which clearly distinguished this intracellular process from known proteolytic enzymes and implied novel biochemical mechanisms. This energy requirement was the critical clue to our finding and elucidation of the responsible degradative system [9], and was at the time most puzzling, because there was clearly no thermodynamic necessity for energy to support peptide bond hydrolysis and no ATP requirement for the function of known proteases. Although studies in mammalian cells in the 1960s and early 1970s, especially by Schimke, Tomkins, and co-workers, had indicated that many tightly regulated enzymes were rapidly degraded in vivo [4], this process ceased when cells were broken open largely because of this energy requirement for intracellular proteolysis [5]. It had been speculated that energy might be required for lysosomal function; however, our finding of a similar ATP requirement in bacteria and reticulocytes indicated that another explanation must apply [5]. On this basis, Joseph Etlinger and I searched for such a system in cell-free extracts of reticulocytes and were able to demonstrate a soluble, non-lysosomal (neutral pH optimum) proteolytic system that carried out the selective degradation of abnormal proteins when provided with ATP in a similar fashion to intact cells [9]. The biochemical dissection of this degradative system by Hershko's and our labs were focused on attempts to understand the molecular basis for this mysterious ATP requirement.

A fundamental advance was the discovery by Hershko, Ciechanover, and Rose of the involvement in this process of a small, heat-stable protein, subsequently identified by others to be the polypeptide ubiquitin, which was known to be linked in vivo to certain histones [8, 10, 11]. Their seminal work between 1979 and 1982 established that ATP was necessary for the formation of chains of ubiquitin protein substrates, which marks them for rapid degradation by the 26S proteasome. These workers also identified the three types of enzymes that are involved in the activation (E1), conjugation (E2), and substrate-specific ligation (E3) of the ubiquitin. Mammalian cells appear to contain at least 20–30 different E2s and hundreds of distinct ubiquitination ligases (E3s), which catalyze degradation of different cell proteins. These enzymes thus provide exquisite selectivity to this degradative pathway and hold promise as very attractive drug targets [10, 11].

In parallel studies of protein breakdown in *E. coli* and mitochondria, we discovered a very different explanation of the ATP requirement for proteolysis. The bacteria and these organelles were shown to contain a soluble ATP-dependent proteolytic system for degradation of abnormal proteins. However, in exploring the basis for this requirement, we discovered a new type of proteolytic enzyme that are large complexes (20–100 times the size of typical proteases) and that are both proteases and ATPases [12, 13]. Bacteria and mitochondria lack ubiquitin, and instead, carry out selective protein breakdown using several ATP-dependent proteolytic complexes (in *E. coli* named Lon or La, ClpAP or Ti, ClpXP, HslUV or FtsH). These enzymes hydrolyze ATP and proteins in a linked process and are selective for different kinds of substrates, and are widely studied as models for understanding the role of ATP in the function of the 26S proteasome.

These investigations thus initially suggested two very different explanations for the ATP requirement for proteolysis, ubiquitin ligation in eukaryotes, and ATP-dependent proteases in prokaryotes. However, subsequent work indicated that both mechanisms must also function in mammalian cells and that the degradation of ubiquitinated proteins and certain non-ubiquitinated substrates must be by an ATP-dependent protease [12, 14]. We had found such an ATP-stimulated proteolytic activity very early, but its isolation, characterization, and precise role took many years to clarify [15]. Finally, in 1987, Rechsteiner's [16] and our groups [17] were able to isolate a very large complex that degraded ubiquitin-conjugated proteins in an ATP-dependent process, the complex we subsequently named the 26S proteasome (see Fig. 2). These eubacterial ATP-dependent proteases appear to function through similar ATP-dependent mechanisms as the 26S proteasome. In fact, one of these ATP-dependent proteases, HslUV, and the homologous proteasomes of archaebacteria and actinomyces appear to be the evolutionary ancestors of the eukaryotic 26S proteasome (see below). Thus, proteasomes evolved quite early, before protein breakdown became linked in eukaryotes to ubiquitin conjugation, which clearly provided opportunities for greater selectivity and regulation.

Initially, the very large complex that degrades ubiquitinated proteins appeared to be very different from the 600-kDa proteolytic complex, which we now call the 20S proteasome (20S and 26S refer to their sedimentation rates in an ultracentrifuge) [16, 17]. This 20S structure, containing multiple peptidases, had been discovered in

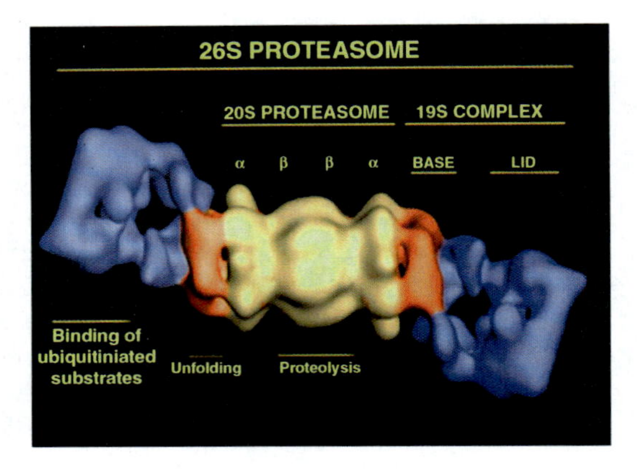

Fig. 2 26S Proteasome and functional roles of its major components. The base of the 19S complex (orange) contains its six ATPase subunits which unfold protein substrates, open the gate in the 20S particle, and translocate proteins into the 20S for deqradation

the early 1980s in several contexts – as a peptidase complex that hydrolyzes neuropeptides by Wilk and Orlowski [18], as a nonproteolytic particle presumed to control translation by Scherrer [12], and as the major ATP-activated cytosolic endoprotease by DeMartino in our laboratory and by others [15]. In the literature, there were actually at least 17 different names and multiple functions proposed for this complex. Eventually, we were able to show that these different structures corresponded to the same particle, which we named the 20S proteasome to indicate that it was a particle with protease function [19]. The next critical step was to demonstrate that the 20S proteasome was essential in the ubiquitin-dependent degradation of proteins in reticulocytes [20]. Finally, we [21] and Hershko's laboratory [22] were able to show that in the presence of ATP, the 20S particle was incorporated into the larger 26S (2.2-kDa) complex that degrades ubiquitin conjugates, which we then named the 26S proteasome.

3 The Rationale for Generating Proteasome Inhibitors for Therapeutic Purposes

These major biochemical advances did not provide any rationale for an effort to synthesize proteasome inhibitors as therapeutic agents. The decision to undertake such an effort came from very different insights, which were also not directly related to cancer or multiple myeloma. In addition to pursuing these biochemical studies, my laboratory continued to carry out physiologic studies on the control of protein breakdown in muscle in normal and disease states. A major advance in this work was the development of simple in vitro techniques for precise measurement of rates of protein breakdown in incubated rodent muscles, which enabled us in the 1970s and 1980s to demonstrate that protein breakdown in muscle was accelerated in various diseases (e.g., cancer cachexia, renal failure, sepsis) in which muscle wasting is prominent (for reviews, see [2] and [3]). Initially, it was assumed that this general acceleration of proteolysis was due to the activation of the lysosomal (autophagic) pathway for proteolysis or the Ca^{2+}-activated proteases (calpains). However, blocking their function in isolated muscles had no effect on the excessive proteolysis, and with time, we were able to demonstrate that the increased protein degradation was primarily due to an activation of the ubiquitin–proteasome pathway (2, 3).

This finding was surprising, since it had been generally assumed that this system primarily degraded the short-lived proteins in cells (e.g., abnormally folded polypeptides or regulatory molecules) [9–11] and not the long-lived proteins (e.g., contractile proteins), which comprise the bulk of cell proteins. In fact, it is now clearly established that atrophying muscles (whether due to fasting, denervation, cancer, sepsis, or diabetes) undergo a common series of transcriptional adaptations that enhance its capacity for proteolysis [2, 3], including increased expression of ubiquitin, proteasomes, and key new ubiquitination enzymes (E3s) [23, 24]. Thus, there is a specific transcriptional program that is activated in various disease states leading to enhanced activity of the ubiquitin–proteasome pathway and rapid destruction of muscle proteins [25–27].

These insights led me to propose that it might be of major benefit to an enormous number of patients to be able to inhibit pharmacologically this degradative pathway in muscle, especially since only a relatively small increase in overall proteolysis (two- to three-fold) appeared to be responsible for the rapid muscle atrophy. Therefore, our initial goal in undertaking to synthesize proteasome inhibitors was to partially inhibit the proteasome and thus to reduce muscle proteolysis to its normal rate in these catabolic states. At the time, there was no mechanism within the university to bring together the enormous expertise that existed there in chemistry, biochemistry, cell biology, mammalian physiology, and medicine – all the skills that one would want to bring to bear to pursue a drug development effort. By creating a biotechnology company, we could marshal such a talented scientific group to function together as a team, which was not how scientific research was or is done in academia. Also, I had personally some very positive experiences in interacting with biotechnology companies: I had been involved in the 1980s in consulting to several of the emerging recombinant DNA companies and was able in this role to experience the excitement and satisfaction in seeing our experience and discoveries about protein breakdown in bacteria contribute to development of new therapies.

4 The Failed (But Scientifically Very Successful) Biotechnology Company, Myogenics/Proscript

Therefore, in the early 1990s, I convinced a group of Harvard colleagues to form a Scientific Advisory Board and to help found a biotechnology company whose primary goal would be to try to control the debilitating loss of muscle in these diseases by retarding the ubiquitin–proteasome pathway. Founding a company was attractive, because in a university setting, in which research programs are restricted by individuals' grants, it is really impossible to bring together faculty with diverse expertise to work as a team toward a common goal. Eventually, we were able to obtain a major investor who had the willingness to gamble on this novel disease target (muscle wasting and cachexia) and on this poorly understood biochemical target, the ubiquitin–proteasome pathway.

A small biotechnology company was founded in Cambridge, Massachusetts in 1993. It was initially named MyoGenics to indicate our goal of preventing the debilitating loss of muscle, and our first target was to develop inhibitors of the proteasome. We quickly assembled a very talented, small group: enzymologists led by Ross Stein, medicinal chemists led by Julian Adams, and cell biologists led by Vito Palombella, whose efforts led eventually to the synthesis and preclinical development within several years of the proteasome inhibitor bortezomib.

In many ways, MyoGenics (later renamed ProScript) was a very unusual biotech company, and although it was short lived as a business entity, it was scientifically exceptionally successful, both in its own basic discoveries as well as in drug development. One way this company was unusual was in its dependence on basic research and close collaborations with academia; in fact, its scientific board met almost monthly with scientists from the company, and the early scientific knowledge developed rapidly through fruitful collaborations between several of us Harvard-based founder-scientists and the talented enzymologists and biochemists in the company. For example, as soon as the first inhibitors were available, their effects on muscle cells were analyzed [28] in my laboratory and on antigen presentation in Kenneth Rock's (then at the Dana-Farber Cancer Institute) [28, 29], and subsequent studies showing effects on nuclear factor-κB (NFκB) were pursued in collaboration between company scientists and Tom Maniatis's laboratory at Harvard [30]. Also, company scientists made many fundamental findings about their pathway and important methodological advances. For example, they obtained the first evidence for the regulatory role of ubiquitination outside the proteasome pathway, and for enzyme regulation by the ubiquitin-like modifier, Nedd8. These scientific strengths did not interfere with its ability to also develop drugs; two proteasome inhibitors emerged (bortezomib and a lactocystin-like molecule) that eventually went through Phase II trials and additional drug candidates later emerged from programs they initiated. Unfortunately, the business entity did not last long enough to enjoy their eventual successes.

In addition to this official goal of combating cachexia and muscle atrophy, I had a further agenda, which was kept secret (especially from the investors). Beyond possible therapeutic benefits, I recognized that if we could develop specific inhibitors of the proteasome that could enter cells, we would have very valuable tools to define the physiological functions of the ubiquitin–proteasome system. However, venture capitalists, stockholders, and company executives are not motivated by their interest in advancing biological science. Therefore, this motivation was kept as a hidden agenda of mine, although it has proved to be a major legacy of MyoGenics/ProScript. In fact, both these goals, advancing basic knowledge and addressing unmet medical challenges, were achieved. Our introduction of proteasome inhibitors greatly advanced our understanding of many aspects of cell regulation, disease mechanisms, and immune surveillance [31–33] (see below). In fact, very few biotechnology (or pharmaceutical) companies, especially small ones, have had such a marked effect on an area of science or published so often in major scientific journals. One special but unusual step was that this company also very early distributed some of our first proteasome inhibitors (especially the widely used agent, MG132) freely to academic investigators, whose efforts rapidly advanced our knowledge of their effects. For example, early academic studies with these inhibitors remarkably altered our knowledge of the proteasome's importance in cancer, apoptosis, inflammation, and antigen presentation. At the same time, these studies ruled out as impractical certain potential therapeutic applications, such as the possibility of using them to suppress immune responses.

5 From MG132 to Bortezomib

The first proteasome inhibitors synthesized by the company were simple peptide aldehydes [29, 31–32], which were analogs of the preferred substrates of the proteasome's chymotrypsin-like active site (Fig. 3). Genetic studies in yeast had suggested that this site was the most important one in protein breakdown [14, 34], and we knew that hydrophobic peptides would be likely to penetrate cell membranes readily, especially if their N-termini was blocked. The C-termini of peptide substrates was derivatized to form an aldehyde group, which was known to be an effective inhibitor of serine and cysteine proteases. Subsequent work using X-ray diffraction disclosed that the proteasome actually has a unique mechanism of action (see below). The first inhibitors synthesized at the company were peptide aldehydes and called MG compounds (for MyoGenics), such as MG132, which has now been employed as a research tool in several thousand scientific papers. It is still the proteasome inhibitor most widely used in basic research in cell biology because it is inexpensive, and its actions are readily reversible [31–33]. (Its peptide backbone is in fact only three leucine molecules plus a blocking group on its N-termini to promote cell permeability.) Most other inhibitors subsequently synthesized, including bortezomib, were derived from these initial molecules by medicinal chemistry, in which a peptide with high affinity for the proteasome's chymotrypsin site was linked to different inhibitory pharmacophores such as a vinyl sulfone (MG412) or a boronate (MG262) [33].

In retrospect, the most important finding for drug development that emerged from our early studies at Harvard in collaboration with Ken Rock with the first inhibitors was that reducing or blocking proteasome function in vivo did not immediately kill cells or interfere with their normal functions, both of which were theoretical possibilities [28, 29]. In fact, our great fear in pursuing this target was that inhibition of the proteasome would rapidly lead to an accumulation of short-lived cell proteins in inactive ubiquitin-conjugated forms, and thus these inhibitors would be highly toxic. However, the presence in cells of a large number

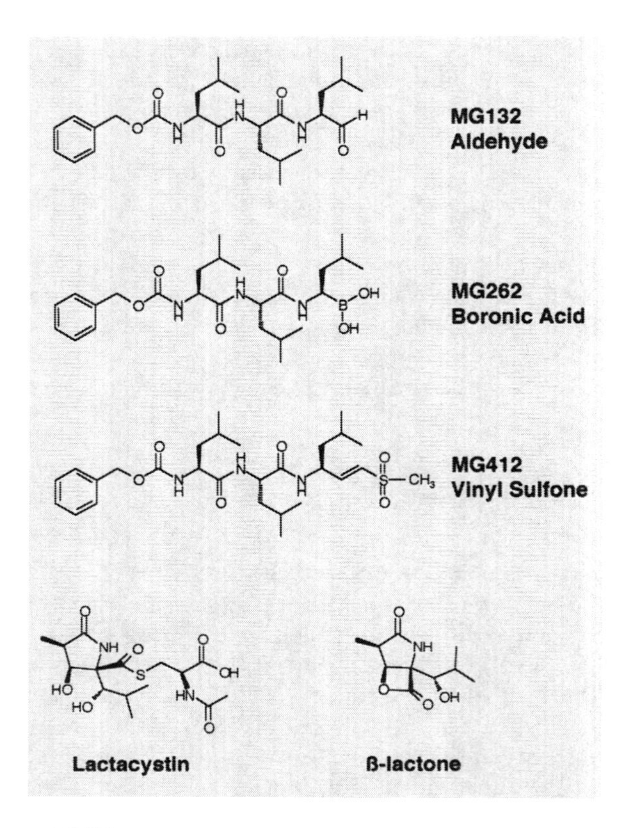

Fig. 3 Proteasome inhibitors

of deubiquitinating enzymes (i.e., DUBs or isopeptidases) that remove ubiquitin from proteins means that only a small fraction of cell proteins accumulate in the ubiquitinated form, even after marked inhibition of proteasomes. This simple observation meant that proteasome inhibitors could be drugs and not just laboratory agents. In other words, cells could function quite well for hours or even days with significantly reduced proteasomal capacity, as became evident from subsequent animal studies and clinical trials.

Julian Adams, who was then recruited to lead our chemistry team, introduced the highly potent boronate group as the inhibitory warhead (hence the name bortezomib). The boronate pharmacophore had originally been developed to inhibit serine-proteases by scientists at DuPont, but this group actually proved to be much more potent and selective as inhibitors of the proteasome. (In fact, the boronate addition to MG132 (Fig. 3) increased inhibitory potency against the proteasome over 50-fold.) Within several months, Adam's team of medicinal chemists modified the peptidic portion to generate bortezomib, a dipeptide boronate, which was initially named MG341, but it has since undergone multiple name changes as the company underwent various transitions (Fig. 4).

The major developments that eventually led to a change in the company's focus came from the original discovery by Vito Palombella [35] when he was a postdoc in Tom Maniatis's laboratory in a collaboration with my laboratory. He showed that the 26S proteasome was critical in the activation of the key transcription factor, NF-κB, which plays a fundamental role in both inflammation and cancer. Eventually, because of NF-κB's central role in the inflammatory response and neoplasia, the company's focus changed to the anti-inflammatory and antineoplastic actions of bortezomib, and the original goal in blocking muscle atrophy became secondary. The company therefore changed its name to ProScript, for "Proteasomes and Transcription." (Hence, PS-341 became the new name for MG341.) As discussed below, proscript eventually encountered major financial challenges and was taken over by a larger biotech company, Leukosite (hence, bortezomib was for a while LDP-341), which was later bought by Millennium Pharmaceuticals (hence MLN-341), who led its successful clinical trials and rechristened the drug bortezomib Velcade for commercial purposes.

It is also noteworthy that these promising inhibitors were initially synthesized based on simple biochemical knowledge of the specificity of the proteasome's active sites through directed medicinal chemistry, and by the use of classical enzyme and intracellular assays. They were not identified through a purely random screening of huge chemical libraries, as is the practice in most drug development efforts. Moreover, at the time, the nature of the proteasome's architecture and its mode of action were unknown [34]. Therefore, no structural information was

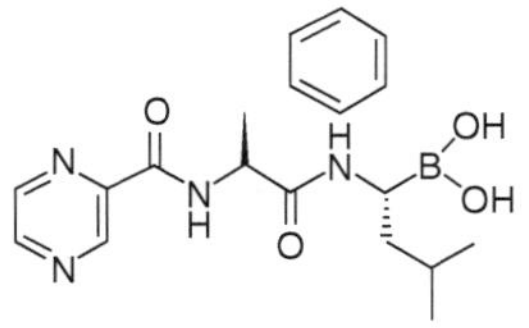

Approved for miltiple myeloma
based on results of phase II trials (2003)

Major mechanisms of anti-neoplastic activity:

Inhibition of NF-kB (anti-apoptotic and generates Il-6 and VEGF)
Causes ER Stress (Unfolded Protein Response) through accumulation of misfolded proteins in secretory pathway & JNK activation
Stabilization of cell-cycle regulatory proteins (p27, p53)
Induction of apoptosis (by JNK and by blocking destruction of caspases by IAPs and can stabilize p53)
Causes accumulation of misfolded proteins in cytosol and nucleus and stress responses (Heat shock response).

Fig. 4 Proteasome inhibitor Bortezomib (Velcade, PS-341)

available from X-ray crystallographic analysis to facilitate drug optimization. Nevertheless, exciting drug candidates emerged within 1 year of the start of this company with less than a dozen talented scientists working on the project.

6 The 20S Proteasome (The Core Particle)

In the nucleus and cytosol of eukaryotic cells, the degradation of most proteins is catalyzed by the 26S proteasome, an exceptionally large, 60 subunit, ATP-dependent proteolytic complex that differs in many fundamental respects from typical proteolytic enzymes [14, 34] (Fig. 2). The great majority of proteases are small 20- to 40-kDa proteins that cleave their substrate once and then release the two fragments. By contrast, proteasomes are highly processive [36]; i.e., they cut polypeptides at multiple sites and degrade the proteins down to small peptides ranging from 2 to 24 residues in length, with a median size of six to seven residues [37]. The 26S proteasome consists of a cylindrical proteolytic particle, the core 20S (720-kDa) proteasome, in association with one or two 19S (890-kDa) regulatory complexes [14, 34, 38], also termed PA700. These complexes associate with each other in an ATP-dependent process [14, 34]. Although generally pictured as the symmetric (19S-20S-19S) complexes, many 26S proteasomes when isolated are asymmetric, singly capped (19S-20S) complexes, and the relative importance of these forms in protein breakdown in vivo is still unclear. Although the 26S proteasome's subunits are nearly identical in all tissues, there are 30–40 proteins that associate with this structure in substochiometric amounts that vary in different tissues and seem to serve as cofactors or regulators of degradation [39].

Free 20S proteasomes also exist in mammalian cells [40, 41], but when isolated by gentle approaches, they are relatively inactive against peptide or protein substrates [14, 34] because protein and peptide substrates cannot enter the particle. In the absence of the 19S particle, the 20S proteasomes are not ATP dependent and are not able to degrade ubiquitin-conjugated proteins. Thus, they are unlikely to play a major role in intracellular proteolysis, which in vivo is largely an ATP-dependent process [5, 7, 42] and generally requires substrate tagging by linkage to a chain of ubiquitin molecules. Additional forms of the proteasome exist in vivo; for example, the cytokine interferon-γ induces a heptameric ring complex, PA28 (also termed Reg), which enhances peptide entry and exit in vivo [43, 45]. Also, single-capped 26S particles may associate with the PA28 (11S) complex [40] to form hybrid 19S-20S-PA28 complexes [44]. These structures enhance the production of MHC class I-presented antigenic peptides by proteasomes [45] in most cells during inflammation and are found constitutively in immune tissues [46, 47].

The 20S proteasome is a hollow, cylindrical particle consisting of four stacked rings. Each outer ring contains seven distinct but homologous α-subunits; each inner ring contains seven distinct but homologous β-subunits [14, 34, 48]. The active sites face the inner chamber of the cylinder. These 20S particles contain three types of active sites encoded in different subunits found on its central ring (Fig. 5).

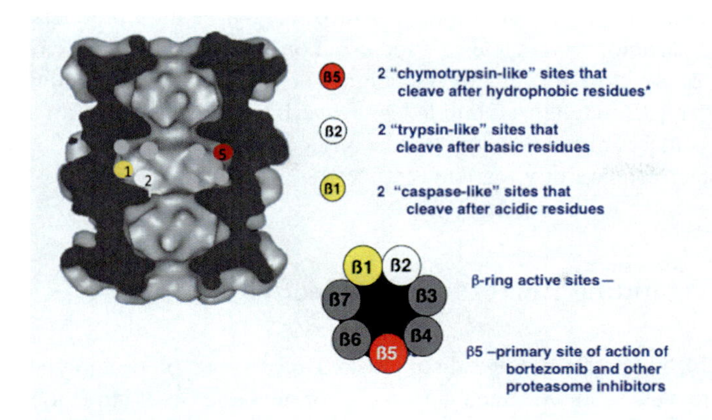

Fig. 5 20S Proteasome has three types of active sites

In each, there is a chymotrypsin-like activity on β5, a trypsin-like site on β1, and a caspase-like site on β2. These sites act synergistically to cleave multiple bonds in proteins (see below). The catalytic N-terminal threonine residue becomes exposed by proteolytic processing of a precursor during proteasome assembly. These sites act synergistically to cleave multiple bonds in proteins [63].

Because the outer walls of the proteasome are very tightly packed, the only way for substrates to reach this central degradative chamber or for products to exit is by passage through a narrow gated channel in the center of the α-rings [49, 50]. This channel is too narrow to be traversed by tightly folded globular proteins; therefore, the breakdown of most proteins requires their unfolding prior to translocation into the core particle [51]. This entry channel in the α-ring is tightly regulated [52–55]. This gate prevents the nonspecific entry of cellular proteins and is formed by the N-termini of the seven α-subunits. In eukaryotes, the backbone of the gate is the N-terminus of the α-3 subunits with which the others interact. A major role of the ATPases [52, 53] and the binding of ubiquitinated proteins [54] is to facilitate the entry process by promoting gate opening when an appropriate ubiquitinated substrate binds to the proteasome [54] (see below).

It has long been recognized that 20S proteasomes can be isolated in an active or an inactive (latent) form, which can be activated by various treatments (e.g., low concentrations of detergents) [14]. In the inactive 20S proteasomes, these entry channels are closed, as demonstrated by X-ray diffraction [49], whereas the active forms all have open gates, allowing substrate entry. Spontaneous gate opening and activation are inhibited by intracellular concentrations of potassium [55], and a key function of the 19S (PA700 complex) is to facilitate substrate entry. Binding of ATP by this particle can trigger gate opening as part of the ATP-dependent translocation of substrates into the 20S particle [52, 53, 55]. The diameter of this gate in the α-ring also influences the sizes of peptide products generated during proteolysis [55]. In addition, small hydrophobic peptides allosterically trigger gate opening, which may represent a mechanism by which peptide products exit [56]. Because the particle's active sites are localized on its inner cavity, this architecture must have evolved to prevent the uncontrolled destruction of cellular proteins. Similarly, the 19S regulatory complexes, as well as substrate ubiquitination, may be viewed as mechanisms that ensure the entry of substrates into the 20S particle is a highly selective, carefully regulated process.

7 Understanding Bortezomib's Selectivity

Traditionally, biochemists have distinguished four types of proteolytic enzymes according to their catalytic mechanisms – serine proteases (e.g., blood clotting enzymes), cysteine proteases (e.g., caspases), metalloproteases, and acid proteases (like HIV protease). However, proteasomes were found to comprise a new class of proteolytic enzymes called threonine proteases, whose catalytic mechanism differs from that of proteases known previously [32, 34]. The active sites in proteasomes utilize the hydroxyl group of the N-terminal threonine residues on three β-subunits as the nucleophile that attacks peptide bonds. The proteasome thus is an "N- terminal hydrolase," a family of enzymes that have similar three-dimensional structures [57]. Much of our understanding of this unique proteolytic mechanism has come through studies using proteasome inhibitors [32] and site-directed mutagenesis of yeast or archaeal proteasomes [14, 34]. The first evidence that the threonine hydroxyl is the catalytic nucleophile was the finding by X-ray diffraction that a peptide aldehyde inhibitor (ALLN) forms a hemiacetal bond with the hydroxyl group of the N-terminal threonines of the proteasome's β-subunits [49, 58]. Also, mutation of this threonine to alanine completely abolished the activity of the proteasome, whereas mutation to a serine retained significant activity against small peptides [59]. This catalytic threonine residue is covalently modified by the irreversible proteasome inhibitors, the vinyl sulfones or the highly specific natural product inhibitors, lactacystin [60, 61], and epoxyketones [62], but bortezomib and the peptide aldehyde inhibitors (MG132) form a reversible complex with this threonine hydroxyl group that mimics the transition state intermediate during peptide bond cleavage (Fig. 6).

The proteasome thus lacks the catalytic triad characteristic of serine and cysteine proteases [32], which has facilitated the development of proteasome-specific inhibitors. Instead, the free N-terminal amino group of catalytic threonine is likely to accept the proton from the side chain hydroxyl. Since all protein translation at the ribosome begins with an N-terminal methionine, this catalytic N-terminal threonine has to be generated proteolytically during proteasome assembly. In this process, two smaller inactive precursors come together, and a proptide is autolytically cleaved off the three active sites containing subunits exposing the critical threonine [62a]. To summarize the proteasome's catalytic mechanism: First, the hydroxyl group of the catalytic threonine directly attacks the scissile bond, resulting in the formation of the tetrahedral intermediate, which then collapses into an acyl enzyme

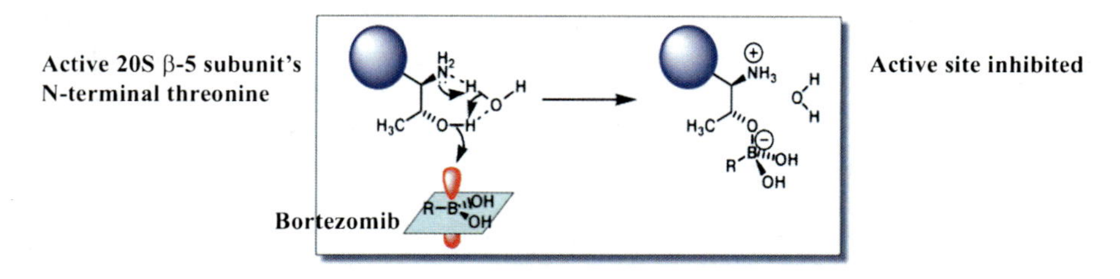

Fig. 6 Mechanism of proteasome inhibition by peptide boronates (e.g., Bortezomib)

with the release of the first reaction product. Deacylation of the catalytic threonine residue by water leads to the release of the second peptide product and the regeneration of the free N-terminal threonine on the proteasome's active site.

A functional proteasome is absolutely essential for life, and cells with deletion mutations in any of the genes for its core subunits are unable to grow. Not surprisingly, many individuals familiar with the importance of the proteasome pathway had assumed that inhibitors of proteasome function would be highly toxic. Nevertheless, as our early studies in animals and the experience of thousands of patients indicates that Bortezomib is not very toxic at therapeutic doses. Studies by Alexei Kisselev in our laboratory [63] on pure proteasomes have helped explain this difference. These studies were undertaken to clarify the contributions of each of the proteasome's three types of active site to the breakdown of different model proteins. By selectively inactivating each type in purified proteasomes, Kisselev showed that all of them contribute to cutting of proteins, but no single type is absolutely essential. In other words, the relative contributions of each type of active site depend upon the sequence of the protein substrate [63].

Bortezomib and the other proteasome inhibitors commonly used in the laboratory all block primarily the chymotrypsin-like site, which appears to be the most active one in proteolysis. Nevertheless, its inactivation still allows significant but slower protein degradation by the other sites. To block the proteasome's capacity to degrade proteins, as scientists do in cultured cells, these inhibitors must be used at higher concentrations, where they block two types of sites. For example, bortezomib at higher concentrations then also inhibits the caspase-like site, but even when two sites are inhibited, the remaining sites are capable of degrading proteins at 20–30% of its usual rate. Most importantly, these quantitative studies indicated that at therapeutic doses seen in patients, bortezomib only partially inhibits protein degradation, probably by only 20–30% [63]. While most cells of the body are unaffected, the myeloma cells appear particularly susceptible to this degree of inhibition, presumably due to their very high rates of synthesis and breakdown of abnormal immunoglobulins, and even mild inhibitions can push them into apoptosis.

The inhibitors of the proteasome available now all inhibit the active sites of the 20S particle, primarily the chymotrypsin-like site. However, as noted below, the 26S particle contains many subunits and multiple enzymatic activities in its 19S component. Therefore, it seems very likely that this particle contains many other possible targets for selective inhibition and perhaps drug development in the future. At least in theory, agents affecting the 19S particle might even be anticipated to affect degradation of different substrates of the proteasome selectively and thus might prove more useful for certain diseases than inhibitors now available.

8 The 19S Regulatory Particle and the Role of ATP in Proteasome Function

The other component of the 26S proteasome is the 19S regulatory complex (PA700), and in recent years, appreciable progress has been made in defining its composition and the functions of its individual subunits [11, 34, 39] (Fig. 2). The 19S particle can be separated into a base and a lid [11, 64]. The lid contains at least nine polypeptides, including multiple isopeptidases that disassemble the polyubiquitin chain, allowing free ubiquitin to be reutilized in further rounds of proteolysis. The removal of the ubiquitin chain is an ATP-dependent process catalyzed by a specific subunit (Rpn11), and this step is essential for ATP-dependent degradation of the substrate [65, 66] by the 20S particle. The base, which associates with the 20S particle, contains eight polypeptides, including six homologous ATPases, which serve multiple functions. They interact directly with the α-rings of the 20S to trigger ATP-dependent opening of the channel, which is essential for polypeptide entry into the proteolytic chamber [52, 53]. The ATPases also have chaperone-like functions that enable them to bind unfolded polypeptide substrates, trigger gate opening, unfold a globular protein, and catalyze protein translocation into the 20S proteasome [52, 67] (Table 1).

Much has been learned about the role of ATP hydrolysis by studying the analogous complexes from prokaryotes, especially archaebacteria, which lack ubiquitin but contain simpler forms of the 20S proteasomes [68]. These particles function in protein breakdown together with the hexameric ATPase ring complex, which we named PAN (for "Proteasome Activating Nucleotidase") [69, 70]. PAN shares more than 40% identity with the six ATPases in the base of the 19S

Table 1 ATP stimulates multiple steps in protein degradation by the 26S proteasome that precede substrate entry into the 20S

Enables the 26S proteasome to bind tightly ubiquitinated proteins
Catalyzes the unfolding of globular proteins (This is the one step that absolutely requires the energy from ATP hydrolysis)
Facilitates the diffusion of unfolded protein substrate into the 20S core particle
Opens the gates for substrate entry into the 20S particle
Maintains the association of the 29S regulatory particle with the 20S proteasome

Conclusion. The proteasome is an ATP-driven proteolytic machine whose functioning is orchestrated by the six ATPases found in the base of the 19S regulatory complex

complex and thus appears to be the evolutionary precursor to the 19S base, which must have regulated proteasome function in archaea before proteolysis became linked to ubiquitination in eukaryotes.

The proteasome regulatory ATPases are all members of the large AAA family of hexameric ATPases [71], which includes several enzymes that we had discovered in the 1980s or 1990s and showed to be critical in intracellular protein degradation in *E. coli* and mitochondria [12, 72]: ATP-dependent protease Lon and the regulatory components of the bacterial ATP-dependent proteases, ClpAP, ClpXP, and HslUV [72], as well as the eukaryotic p97/VCP complex (a key factor in the degradation of many ubiquitinated proteins [73, 74]. Like these enzymes, PAN (and presumably the 19S ATPases) increases its rate of ATP consumption several fold when it binds an appropriate substrate [67]. This ATPase complex has been shown to catalyze ATP-dependent unfolding of the globular protein, GFPssrA [75], which when unfolded loses its green color. This process seems to involve in ATP-driven threading the polypeptide through the ATPase ring and passage through the narrow central opening in the ring [76] seems to unfold the upstream domains, but its detailed mechanism is still unclear. In addition, the ATPase complex is necessary for the rapid entrance of proteins, even denatured ones, into the core proteasome (Table 1). These substrates appear to be translocated through a central opening in the ATPase ring and then through the gate in the α-ring [76]. Some proteins enter exclusively in a C to N direction, whereas others are translocated in

Table 2 Key functions of the proteasome discovered or elucidated using proteasome inhibitors as research tools

Proteasomes are the primary site for protein degradation in mammalian cells [29] – they catalyze the degradation of most long-lived proteins in cells (i.e., the bulk of cell proteins) as well as the short-lived, rapidly degraded components (i.e., regulatory proteins or misfolded polypeptides)
Proteasomes are the source of most of the antigenic peptides presented on MHC-class 1 molecules [29]; thus, the immune system uses proteasome products as the basis for surveillance for viruses, intracellular pathogens, and cancer, which elicit cytotoxic T-cell responses
Endoplasmic Reticulum (ER)-Associated Degradation (the ERAD Pathway) – misfolded or mutated secretory or membrane proteins are selectively eliminated by extraction from the ER and degradation by cytosolic proteasomes [73]. This quality control process is linked to the "Unfolded Protein Response," the stress response induced by accumulation of abnormal proteins within the ER. It is particularly important in myeloma cells
At normal oxygen tensions, cells rapidly degrade the HIF-1α transcription factor, which prevents the transcriptional response to hypoxia. In hypoxia or ischemia, HIF-1α accumulates, leading to activation of genes involved in erythropoiesis, angiogenesis, and glycolysis
Continual proteasome function prevents apoptosis through functioning of the IAP (Inhibitor of Apoptosis Proteins) that prevents activation of caspases and promote their degradation
The clock mechanisms that underlie our diurnal rhythms depend on programmed degradation of key time-setting proteins
Cell growth, proliferation, and metabolism are controlled largely through degradation of short-lived regulatory proteins. Among these rapidly degraded proteins are most Oncogenes and tumor suppressors, transcription factors, cell cycle regulators (cyclins or CDKs), and rate-limiting enzymes

an N to C direction. Several studies have determined the actual amounts of ATP utilized during degradation of model proteins by the PAN proteasome complex [67]. Surprisingly, for the unfolded substrate casein and the tightly folded protein GFPssrA, the same amount of ATP is hydrolyzed, about 350 ATP molecules/ molecule of the protein, which is perhaps a third of what is consumed by the ribosome in synthesis of a polypeptide of this size [67]. Thus, the cell invests significant ATP to insure that the degradation of protein is a highly efficient and highly selective process.

To clarify the roles of ATP in the function of the 19S proteasome-associated ATPases in recent years, we have studied in depth the much simpler, but homologous archaeal 20S proteasome, which offer many advantages for mechanistic studies. As in mammalian particles, substrates enter archaeal proteasomes through a narrow gated channel in its outer α-ring and the N-termini of the α-subunits function as an ATPase-regulated gate that in its closed form prevents nonspecific entry of protein substrates. ATP binding allows gate opening, but ATP hydrolysis to ADP restores the closed form [52, 53]. Thus, ATP binding alone can activate multiple key steps in proteasome function (complex formation, gate opening, and translocation of unfolded proteins), but degradation of globular proteins requires energy-dependent unfolding (Table 1).

Recently, we have succeeded in elucidating this mechanism by which gate opening occurs [52, 53, 77, 78]. These ATPases contain a C-terminal tripeptide (HbYX) motif, which bind to pockets in the 20S's α-ring and function like "keys-in-a-lock" to trigger gate opening. Short peptides from the ATPases C-termini by themselves can trigger gate opening, provided they contain the HbYX motif. Using single-particle cryo-EM and X-ray diffractions, we showed that the C-termini interact with key residues in the inter-subunit pockets and cause gate opening by inducing slight rotation of the α-subunits [77, 78]. In eukaryotes, peptides from only two ATPases, Rpt2 and 5, can trigger gate opening, and the termini of the other ATPases serve to insure the tight association of the 19S with the 20S particle [53]. Thus, the C-termini of the six ATPases in eukaryotes all dock into pockets in the 20S particle, but seem to have acquired specialized functions, and our latest studies indicate that the six bind and cleave ATP in a highly ordered fashion to drive proteolysis.

To understand the selectivity of the 26S proteasome for ubiquitinated proteins, we tested if the ubiquitin conjugates might also activate the proteasome's capacity for proteolysis [54]. After binding a polyubiquitinated protein, mammalian 26S proteasomes hydrolyzed peptide substrates 2- to 7-fold faster. The Ub conjugates enhanced peptide hydrolysis by stimulating gate opening in the 20S, and this effect required nucleotide binding. To cause gate opening, conjugates associate with the 19S deubiquitinating enzyme subunit, Usp14/Ubp6 [54]. No stimulation was observed with 26S lacking this enzyme (from Ubp6Δ mutants), but was restored by addition of pure Ubp6. Thus, conjugate binding to Usp14/Ubp6 regulates substrate entry into the 20S particle. This mechanism must help enhance the selectivity of the 26S proteasome for ubiquitinated proteins and links ubiquitin chain disassembly to protein degradation. In other words, the ubiquitin chain is not just a marker for destruction, but it helps regulate the function of the 26S proteasome, which should be viewed as an intricate multifunctional molecular machine.

9 Bortezomib's Preclinical Trials and Tribulations

Compared to the great majority of drugs approved in recent years, bortezomib's development was unusually rapid. However, its path from test tube through preclinical development to human trials was hardly direct or easily. At multiple junctures, the entire program had to be cut back and came close to termination, primarily for non-scientific (i.e., financial) reasons. Progress was initially very swift. Only a few years passed from the first evidence that proteasome inhibitors could reduce protein degradation in intact cells in 1994 [29] to the synthesis of peptide boronates and finally bortezomib by Julian Adams' team to the acquisition of strong evidence for efficacy in cells and mouse models of inflammatory disease and cancer through work led by Vito Palombello and Peter Elliott. Eventually, supporting data for efficacy in cancer models came from screening efforts at the National Cancer Institute and academic investigators and led to NCI's commitment to support clinical development of bortezomib.

This rapid progress may suggest a problem-free development program. On the contrary, a plethora of disconnected challenges was encountered that delayed, and almost derailed, these efforts several times. Most of these problems were related to the development of ProScript as a business entity, whose administrative leadership clearly did not excel as did its scientific programs. In fact, they led to the early demise of our company ProScript just when the path to the clinic was imminent. Among the variety of unforeseen challenges was major weakness at the time in the biotechnology industry leading to greatly decreased investment internationally, which was complicated by the premature, sudden death of the principal figure in the venture capital firm that had been the sole investor in ProScript. More critically, we had established a promising partnership with Hoechst Pharmaceuticals in which bortezomib was initially investigated as an anti-inflammatory agent, because of its ability to inhibit NF-κB activation and its beneficial effects in several experimental models of inflammatory disease (i.e., rheumatoid arthritis, psoriasis). However, these efforts were unexpectedly terminated when senior management at Hoechst decided to bring to the clinic a different anti-inflammatory developed by their internal staff, and bortezomib was then pursued as an anticancer agent (our backup plan). Despite the highly encouraging data against cancer models, Hoechst's administration subsequently made the strategic decision to terminate all its efforts in the cancer area, which left bortezomib and ProScript's future in limbo.

Eventually, the legal rights to this molecule reverted back to ProScript, but the termination of its relationship with a major pharmaceutical company and Hoechst's decision to not develop bortezomib made obtaining further investment particularly difficult. As its various efforts to obtain new investment were unsuccessful, the company decreased in size, and its programs were cut back on several occasions.

Despite impressive inhibition of many cancers in mouse xenografts, approaches to all the major pharmaceutical companies and large biotechnology companies elicited no interest in partnering or developing bortezomib. At the time, the ubiquitin–proteasome pathway was poorly understood in the pharmaceutical industry, and was not at all considered a possible drug target. Also nearly all experts consulted by potential partners predicted incorrectly that its inhibition would be highly toxic (or that peptide boronates would be harmful), even though prolonged treatment of mice with bortezomib was clearly proving much less toxic than standard chemotherapy.

As a consequence of this inability to obtain new funding, our primary investor, Healthcare Ventures, decided to close ProScript and sell its resources. Eventually, its Board of Directors sold ProScript to a larger biotech company owned by the same venture group, Leukocyte. As a result, the remaining team of ProScript scientists working on bortezomib and the ubiquitin–proteasome system were incorporated into that company largely intact and continued to work on bortezomib's actions and in vivo pharmacology. At the time of the sale, the value of ProScript's assets including rights to other peptide boronates, as well as another proteasome inhibitors, (e.g., a derivative of lactacystin that went as far as Phase II trials for stroke) [54b], and another research program that eventually generated the inhibitor of Neddylation, which recently entered clinical trials plus related intellectual property, was remarkably small (approximately three million dollars). This meagre valuation reflected the lack of interest in proteasome inhibition as a therapeutic strategy. It is noteworthy that this year's (2010) sales of bortezomib are well over 400 times the cost of the entire company.

Ironically, several months after ProScript was purchased, Leukocyte was in turn purchased by the much larger company, Millennium, Inc, whose initial focus had been on genomic approaches to drug development, but had failed to generate drug candidates. The purchase of Leukocyte was quite costly, but it brought to Millennium a pipeline of drug candidates. However, none of the agents led to successful drugs, except for bortezomib, which Millennium also had not valued highly. In fact, at the time Millennium publicly heralded the eight drug programs that it had acquired and failed to even mention the proteasome inhibitors, the one program which was to prove highly successful in the clinic, and the one that led to Millennium's survival and growth as an independent company.

Through all these troubled times, the continuous devotion of the same scientists led by Julian Adams and also the pharmacologist, Peter Elliott, continued to generate evidence of bortezomib's promise. Within Millennium, Julian was a forceful and effective advocate for these programs. Also, it was fortunate for bortezomib's emergence that the other programs being pursued in Millennium faltered with time, leading to their greater focus on bortezomib and its entry into the clinic and its eventual approval against hematological malignancies. Others can tell the history of bortezomib's successful clinical development by Millennium and its academic collaborators much better than I, since I was no longer directly involved in this story (only as a very interested "grandparent" would be in his grandchildren's fate). But its clinical development is also a tale of serendipity.

Eventually, when bortezomib entered Phase I trials against all cancers, one patient showed a dramatic, complete remission. That patient had late-stage multiple myeloma, where there was no precedent for such marked improvement. Some additional responses were then observed in other patients with multiple myeloma. Consequently, Phase II trials focused on this disease, for which there was no adequate therapy.

The Phase II trials were carried out in expert fashion by the team of Ken Anderson and Paul Richardson of the Dana-Farber Cancer Institute (see other chapters in this volume). These trials were a model of thoughtful experimental design, and were aided by the development by the company of sophisticated assays of enzyme inhibition for use in patients, and led to bortezomib approval after only Phase II results. In retrospect, it is ironic that during our preclinical development on cancer, multiple myeloma was not even considered as a prime therapeutic target, even though the very important role of NF-κB in plasma cells and the role of the proteasomes in quality control of secreted proteins were well established. So despite our scientific sophistication, we failed to foresee the exact impact of our efforts. Beyond the intelligence, creativity, and hard work of many, happenstance clearly played a major role in its emergence. Because of the financial challenge and because of strong concerns about potential toxicity that proved unwarranted. Bortezomib almost got lost or irretrievably buried long before its success in the clinic could be established. This story is probably not unique. It seems likely that other valuable drugs may have been sidelined for the lack of talented advocates, lack of sufficient investment, or poor design of clinical trials, and the vagaries of chance, and their potential for helping humans patients was never realized.

10 The Legacy of the Proteasome Inhibitors Outside the Clinic: Major Advances in Our Understanding of Cell Function

In addition to their major impact on hematology and oncology and the enormous benefits to suffering patients, the availability of selective proteasome inhibitors (primarily the peptide aldehyde inhibitors synthesized by ProScript and to a lesser extent the natural products shown to inhibit the proteasome, lactacystin, and epoximicin) have allowed dramatic advances in our understanding of cell regulation, immune responses, as well as disease mechanisms, especially those contributing to cancer. (In fact, at a recent 5-day FASEB conference on this pathway, that I attended, over two thirds of the presentations used these inhibitors as critical tools to establish their conclusions.)

Prior to the development of proteasome inhibitors, the functions of the ubiquitin–proteasome pathway and its different cellular roles were studied primarily by biochemical methods or by genetic analysis of yeast mutants defective in this process. The degradation of a model protein was typically studied using cell-free extracts (especially from mammalian reticulocytes or, more recently, frog oocytes).

These approaches, unfortunately, are technically difficult, and genetic analysis can be quite time-consuming. Also, many complex cellular processes involving the proteasome have to this day never been reconstituted in cell extracts and cannot be studied in yeast (e.g., antigen presentation or muscle atrophy). The availability since 1994 of specific inhibitors of the proteasome that enter intact cells and block or reduce its function [29] has allowed much more rapid analysis of the role of the proteasome in the breakdown of specific proteins and complex cellular responses [31–33]. Thus, if such inhibitors prevent a decrease in activity of an enzyme or cause an increase in the cellular content of a protein, then proteasome-mediated degradation is very likely to play a key role, especially if these inhibitors cause the protein to accumulate in a ubiquitin-conjugated, high-molecular weight form. However, further biochemical analysis of the process is still necessary to identify the responsible ubiquitination enzymes and the critical regulatory factors (e.g., kinases that may trigger ubiquitination and proteasomal degradation).

Since several thousand studies have now been published that utilized proteasome inhibitors to analyze biochemical or cellular phenomena, it is clearly impossible to summarize the many important insights that emerged from their use. In fact, certain major new areas of research have developed based upon initial studies on the effects of these agents on cellular or immune responses. Therefore, I have listed in Table 2, some of the most fundamental discoveries and most active areas of research made possible in this way.

11 Some Lessons About Drug Development Learned from Bortezomib's Success

Beyond its impact on cancer therapy (the subject of other chapters in this book), bortezomib's development illustrates several key lessons that merit wider appreciation in both the medical and scientific communities as well as by the broader public:

1. Bortezomib illustrates the truth often forgotten by granting agencies and patients that improved therapies, especially in the cancer area, are tightly linked to and rely on advances in our understanding of basic biochemistry and cell biology. In this regard, major credit thus should go to the NIH that over the years has funded the enormous growth of basic knowledge about protein degradation that made proteasome inhibition a therapeutic possibility, and also to its component, the National Cancer Institute, which at a crucial time in bortezomib's development screened for possible applications of proteasome inhibitors and recognized their therapeutic potential.
2. In fact, major advances often emerge from outside established lines of research. It is noteworthy that proteasome inhibition was not being considered by any investigators as a target for cancer treatment, and its novelty certainly slowed its acceptance. Clearly, due to the success of bortezomib, several other proteasome

inhibitors are now in clinical trials. However, in the ubiquitin–proteasome pathway, there are over a thousand other protein components, many of which play crucial roles in human disease, and there exist many other opportunities for drug development, which hopefully will be pursued in coming years.

3. The rapid synthesis and preclinical development of bortezomib was made possible by the concerted efforts of talented scientists in an unusual small biotechnology company with extensive input from academic experts and later clinical investigators. Such teamwork is generally not possible in university settings and clearly should be fostered. For "pure" scientists, bortezomib's success illustrates that basic studies and applied disease-oriented research can be linked successfully. One does not necessarily prostitute oneself in trying to develop an agent that might be a commercial success and improve medical practice. Conversely, tremendous advances in understanding cell regulation and the mechanisms of disease have been made possible by investigators using the proteasome inhibitors as research tools.
4. The paths to apply scientific advances and their benefits are often unpredictable. I certainly never anticipated that in studying or the selective degradation of misfolded proteins in bacteria that this work might be useful in protein production in the recombinant DNA industry. I also never dreamed that investigating the mechanisms of muscle wasting might someday lead to the discovery of proteasomal mechanisms, or that progress in this area would in turn lead to novel insights about antigen presentation or contribute to novel therapies for multiple myeloma. In fact, had I even suggested in a grant proposal that pursuing this research program might eventually have such benefits, the NIH Study Section or the reviewers would have rejected such statements as fantasy, nonsense, or pure hogwash (as I would have done, were I the reviewer).

The correct justification for such studies is that basic knowledge is enabling and that our research was addressing important problems about living systems and disease mechanisms. Eventually, gaining such knowledge opened up opportunities for practical benefits because we kept our eyes open to this possibility.

Acknowledgments The author is grateful to Mary Dethavong for her expert assistance in the preparation of this manuscript (and our other recent articles). The research from Dr. Goldberg's laboratory reviewed here has been supported by grants from the NIH (NIGMS and NIA), the Muscular Dystrophy Association, the Ellison Foundation, and the Multiple Myeloma Foundation. Dr. Goldberg receives royalty income from Millennium, Inc. He also has received special considerations and help from his wife, Dr. Joan Goldberg, a practicing hematologist, who thanks to bortezomib is now convinced that working on the proteasome is a worthwhile activity.

References

1. Goldberg AL (1969) Protein turnover in skeletal muscle II: effects of denervation and cortisone on protein catabolism in skeletal muscle. J Biol Chem 244:3223–3229
2. Mitch WE, Goldberg AL (1996) Mechanisms of muscle wasting. The role of the ubiquitin-proteasome pathway. N Engl J Med 335(25):1897–1905

3. Lecker SH et al (1999) Muscle protein breakdown and the critical role of the ubiquitin-proteasome pathway in normal and disease states. J Nutr 129(1S Suppl):227S–237S
4. Goldberg AL, Dice JF (1974) Intracellular protein degradation in mammalian and bacterial cells. Annu Rev Biochem 43:835–869
5. Goldberg AL, St John AC (1976) Intracellular protein degradation in mammalian and bacterial cells: part 2. Annu Rev Biochem 45:747–803
6. Goldberg AL (1971) A role of aminoacyl-tRNA in the regulation of protein breakdown in *Escherichia coli*. Proc Natl Acad Sci USA 68:362–366
7. Goldberg AL (1972) Degradation of abnormal proteins in *E. coli*. Proc Natl Acad Sci USA 69:422–426
8. Goldberg AL (2005) Nobel committee tags ubiquitin for distinction. Neuron 45(3):339–344
9. Etlinger JD, Goldberg AL (1977) A soluble ATP-dependent proteolytic system responsible for the degradation of abnormal proteins in reticulocytes. Proc Natl Acad Sci USA 74(1):54–58
10. Hershko A, Ciechanover A (1998) The ubiquitin system. Annu Rev Biochem 67:425–479
11. Glickman M, Ciechanover A (2002) The ubiquitin-proteasome proteolytic pathway: destruction for the sake of construction. Physiol Rev 82:373–428
12. Goldberg AL (1992) The mechanism and functions of ATP-dependent proteases in bacterial and animal cells. Eur J Biochem 203(1–2):9–23
13. Wickner S, Maurizi MR, Gottesman S (1999) Posttranslational quality control: folding, refolding, and degrading proteins. Science 286(5446):1888–1893
14. Coux O, Tanaka K, Goldberg AL (1996) Structure and functions of the 20S and 26S proteasomes. Annu Rev Biochem 65:801–847
14a. Tanaka K, Waxman L, Goldberg AL (2005) ATP serves two distinct roles in protein degradation in reticulocytes, one requiring and one independent of ubiquitin. J Cell Biol 96(6):1580–1585
15. DeMartino GN, Goldberg AL (1979) Identification and partial purification of an ATP-stimulated alkaline protease in rat liver. J Biol Chem 254:3712–3715
16. Hough R, Pratt G, Rechsteiner M (1987) Purification of two high molecular weight proteases from rabbit reticulocyte lysate. J Biol Chem 261:2400–2408
17. Waxman L, Fagan JM, Goldberg AL (1987) Demonstration of two distinct high molecular weight proteases in rabbit reticulocytes, one of which degrades ubiquitin conjugates. J Biol Chem 262(6):2451–2457
18. Orlowski M (1990) The multicatalytic proteinase complex, a major extralysosomal proteolytic system. Biochemistry 29(45):10289–10297
19. Arrigo A-P et al (1988) Identity of the 19S 'prosome' particle with the large multifunctional protease complex of mammalian cells (the proteasome). Nature 331:192–194
20. Matthews W et al (1989) Involvement of the proteasome in various degradative processes in mammalian cells. Proc Natl Acad Sci USA 86(8):2597–2601
21. Eytan E et al (1989) ATP-dependent incorporation of 20S protease into the 26S complex that degrades proteins conjugated to ubiquitin. Proc Natl Acad Sci USA 86:7751–7755
22. Driscoll J, Goldberg AL (1990) The proteasome (multicatalytic protease) is a component of the 1500 kDa proteolytic complex which degrades ubiquitin-conjugated proteins. J Biol Chem 265:4789–4792
23. Bodine SC et al (2001) Identification of ubiquitin ligases required for skeletal muscle atrophy. Science 294:1704–1708
24. Gomes M et al (2001) Atrogin-1, a muscle-specific F-box protein highly expressed during muscle atrophy. Proc Natl Acad Sci USA 98(25):14440–14445
25. Jagoe RT et al (2002) Patterns of gene expression in atrophying skeletal muscles: response to food deprivation. FASEB J 16(13):1697–1712
26. Lecker SH et al (2004) Multiple types of skeletal muscle atrophy involve a common program of changes in gene expression. FASEB J 18(1):39–51
27. Sacheck JM et al (2004) IGF-1 stimulates muscle growth by suppressing protein breakdown and expression of atrophy-related ubiquitin-ligases, atrogin-1 and MuRF1. Am J Physiol Endocrinol Metab 287(4):E591–E601

28. Tawa NE, Odessey R, Goldberg AL (1997) Inhibitors of the proteasome reduce the accelerated proteolysis in atrophying rat skeletal muscles. J Clin Invest 100:197–203
29. Rock KL et al (1994) Inhibitors of the proteasome block the degradation of most cell proteins and the generation of peptides presented on MHC class 1 molecules. Cell 78(5):761–771
30. Silverman N, Maniatis T (2001) NF-kappaB signaling pathways in mammalian and insect innate immunity. Genes Dev 15(18):2321–2342
31. Lee DH, Goldberg AL (1998) Proteasome inhibitors: valuable new tools for cell biologists. Trends Cell Biol 8(10):397–403
32. Kisselev AF, Goldberg AL (2001) Proteasome inhibitors: from research tools to drug candidates. Chem Biol 8(8):739–758
33. Lee DH, Goldberg AL (1999) The proteasome inhibitors and their uses. In: Wolf DH, Hilt W (eds) Proteasomes: The World of Regulatory Proteolysis. Landes Bioscience Publishing Co, Georgetown, TX
34. Voges D, Zwickl P, Baumeister W (1999) The 26S proteasome: a molecular machine designed for controlled proteolysis. Annu Rev Biochem 68:1015–1068
35. Palombella VJ et al (1994) The ubiquitin-proteasome pathway is required for processing the NF-kappa-B1 precursor protein and the activation of NF-kappa-B. Cell 78(5):773–785
36. Akopian TN, Kisselev AF, Goldberg AL (1997) Processive degradation of proteins and other catalytic properties of the proteasome from Thermoplasma acidophilum. J Biol Chem 272(3):1791–1798
37. Kisselev AF et al (1999) The sizes of peptides generated from protein by mammalian 26S and 20S proteasomes: implications for understanding the degradative mechanism and antigen presentation. J Biol Chem 274(6):3363–3371
38. Holzl H et al (2000) The regulatory complex of Drosophila melanogaster 26S proteasomes: subunit composition and localization of a deubiquitylating enzyme. J Cell Biol 150(1): 119–129
39. Besche HC et al (2009) Isolation of mammalian 26S proteasomes and p97/VCP complexes using the ubiquitin-like domain from HHR23B reveals novel proteasome-associated proteins. Biochemistry 48(11):2538–2549
40. Tanahashi N et al (2000) Hybrid proteasomes. Induction by interferon-gamma and contribution to ATP-dependent proteolysis. J Biol Chem 275(19):14336–14345
41. Yang Y et al (1995) In vivo assembly of the proteasomal complexes, implications for antigen processing. J Biol Chem 270(46):27687–27694
42. Gronostajski R, Pardee AB, Goldberg AL (1985) The ATP-dependence of the degradation of short- and long-lived proteins in growing fibroblasts. J Biol Chem 260:3344–3349
43. Whitby FG et al (2000) Structural basis for the activation of 20S proteasomes by 11S regulators. Nature 408(6808):115–120
44. Cascio P et al (2002) Properties of the hybrid form of the 26S proteasome containing both 19S and PA28 complexes. EMBO J 21(11):2636–2645
45. Rechsteiner M, Realini C, Ustrell V (2000) The proteasome activator 11S REG (PA28) and class I antigen presentation. Biochem J 345(1):1–15
46. Rock KL, Goldberg AL (1999) Degradation of cell proteins and the generation of MHC class I-presented peptides. Annu Rev Immunol 17:739–779
47. Goldberg AL et al (2002) The importance of the proteasome and subsequent proteolytic steps in the generation of antigenic peptides. Mol Immunol 1169:1–17
48. Baumeister W et al (1998) The proteasome: paradigm of a self-compartmentalizing protease. Cell 92:367–380
49. Groll M et al (1997) Structure of 20S proteasome from yeast at 2.4 Å resolution. Nature 386(6624):463–471
50. Groll M et al (2000) A gated channel into the proteasome core particle. Nat Struct Biol 7(11): 1062–1067
51. Wenzel T, Baumeister W (1995) Conformational constraints in protein degradation by the 20S proteasome. Nat Struct Biol 2(3):199–204

52. Smith DM et al (2005) ATP binding to PAN or the 26S ATPases causes association with the 20S proteasome, gate opening, and translocation of unfolded proteins. Mol Cell 20(5):687–698
53. Smith DM et al (2007) Docking of the proteasomal ATPases' carboxyl termini in the 20S proteasome's alpha ring opens the gate for substrate entry. Mol Cell 27(5):731–744
54. Peth A, Besche HC, Goldberg AL (2009) Ubiquitinated proteins activate the proteasome by binding to Usp14/Ubp6, which causes 20S gate opening. Mol Cell 36(5):794–804
54a. Goldberg AL, Rock K (2002) Not just research tools–proteasome inhibitors offer therapeutic promise. Nat Med 8(4):338–340
54b. Soucy TA, Smith PG, Milhollen MA, Berger AJ, Gavin JM, Adhikari S, Brownell JE, Burke KE, Cardin DP, Critchley S, Cullis CA, Doucette A, Garnsey JJ, Gaulin JL, Gershman RE, Lublinsky AR, McDonald A, Mizutani H, Narayanan U, Olhava EJ, Peluso S, Rezaei M, Sintchak MD, Talreja T, Thomas MP, Traore T, Vyskocil S, Weatherhead GS, Yu J, Zhang J, Dick LR, Claiborne CF, Rolfe M, Bolen JB, Langston SP (2009) An inhibitor of NEDD8-activating enzyme as a new approach to treat cancer. Nature 458(7239):732–736
55. Köhler A et al (2001) The axial channel of the proteasome core particle is gated by the Rpt2 ATPase and controls both substrate entry and product release. Mol Cell 7(6):1143–1152
56. Kisselev AF, Kaganovich D, Goldberg AL (2002) Binding of hydrophobic peptides to several non-catalytic sites promotes peptide hydrolysis by all active sites for 20S proteasomes. Evidence for peptide-induced channel opening in the alpha-rings. J Biol Chem 277(25): 22260–22270
57. Brannigan JA et al (1995) A protein catalytic framework with an N-terminal nucleophile is capable of self-activation. Nature 378(6555):416–419
58. Löwe J et al (1995) Crystal structure of the 20S proteasome from the archaeon *T. acidophilum* at 3.4 Å resolution. Science 268(5210):533–539
59. Kisselev AF, Songyang Z, Goldberg AL (2000) Why does threonine, and not serine, function as the active site nucleophile in proteasomes? J Biol Chem 275(20):14831–14837
60. Fenteany G et al (1995) Inhibition of proteasome activities and subunit-specific amino-terminal threonine modification by lactacystin. Science 268(5211):726–731
61. Bogyo M et al (1997) Covalent modification of the active site Thr of proteasome beta-subunits and the *E. coli* homologue HslV by a new class of inhibitors. Proc Natl Acad Sci USA 94:6629–6634
62. Meng L et al (1999) Epoxomicin, a potent and selective proteasome inhibitor, exhibits in vivo antiinflammatory activity. Proc Natl Acad Sci USA 96(18):10403–10408
62a. Kaneko T, Hamazaki J, Iemura S, Sasaki K, Furuyama K, Natsume T, Tanaka K, Murata S (2009) Assembly pathway of the Mammalian proteasome base subcomplex is mediated by multiple specific chaperones. Cell 137(5):914–925
62b. Besche HC, Peth A, Goldberg AL (2009) Getting to first base in proteasome assembly. Cell 138(1):25–28
63. Kisselev AF, Callard A, Goldberg AL (2006) Importance of the proteasome's different proteolytic sites and the efficacy of inhibitors varies with the protein substrate. J Biol Chem 281:8582–8590
64. Glickman MH et al (1998) A subcomplex of the proteasome regulatory particle required for ubiquitin-conjugate degradation and related to the COP9-signalosome and eIF3. Cell 94(5): 615–623
65. Hochstrasser M (2002) New proteases in a ubiquitin stew. Science 298:549–552
66. Verma R et al (2002) Role of Rpn11 metalloprotease in deubiquitination and degradation by the 26S proteasome. Science 298(5593):611–615
67. Benaroudj N et al (2003) ATP hydrolysis by the proteasome regulatory complex PAN serves multiple functions in protein degradation. Mol Cell 11(1):69–78
68. Zwickl P, Goldberg AL, Baumeister W (2000) Proteasomes in prokaryotes. In: Wolf DH, Hilt W (eds) Proteasomes: the world of regulatory proteolysis. Landes Bioscience, Georgetown, TX
69. Zwickl P et al (1999) An archaebacterial ATPase, homologous to ATPases in the eukaryotic 26S proteasome, activates protein breakdown by 20S proteasomes. J Biol Chem 274(37): 26008–26014

70. Wilson HL et al (2000) Biochemical and physical properties of the *Methanococcus jannaschii* 20S proteasome and PAN, a homolog of the ATPase (Rpt) subunits of the eucaryal 26S proteasome. J Bacteriol 182(6):1680–1692
71. Ogura T, Wilkinson AJ (2001) AAA+ superfamily ATPases: common structure – diverse function. Genes Cells 6(7):575–597
72. Gottesman S, Maurizi M, Wickner S (1997) Regulatory subunits of energy-dependent proteases. Cell 91:435–438
73. Raasi S, Wolf DH (2007) Ubiquitin receptors and ERAD: a network of pathways to the proteasome. Semin Cell Dev Biol 18(6):780–791
74. Medicherla B, Goldberg AL (2008) Heat shock and oxygen radicals stimulate ubiquitin-dependent degradation mainly of newly synthesized proteins. J Cell Biol 182(4):663–673
75. Benaroudj N, Goldberg AL (2000) PAN, the proteasome activating nucleotidase from archaebacteria, is a molecular chaperone which unfolds protein substrate. Nat Cell Biol 2(11): 833–839
76. Navon A, Goldberg AL (2001) Proteins are unfolded on the surface of the ATPase ring before transport into the proteasome. Mol Cell 8(6):1339–1349
77. Rabl J et al (2008) Mechanism of gate opening in the 20S proteasome by the proteasomal ATPases. Mol Cell 30(3):360–368
78. Yu Y, Smith DM, Kim HM, Rodriguez V, Goldberg AL, Cheng Y (2010) Interactions of PAN's C-termini with archaeal 20S proteasome and implications for the eukaryotic proteasome-ATPase interactions. EMBO J 29(3):692–702

Preclinical Activities of Bortezomib in MM, the Bone Marrow Microenvironment and Pharmacogenomics

Teru Hideshima, Paul G. Richardson, and Kenneth C. Anderson

Abstract The intracellular protein degradation system is critical for many cellular processes, including cell cycle regulation. The proteasomes are intracellular protein complexes that degrade polyubiquitinated proteins. Bortezomib (Velcade®) is a boronic acid dipeptide that directly binds to enzymatic complex to block chimotrypsin-like activity of proteasome and is the first FDA-approved proteasome inhibitor for multiple myeloma (MM) treatment. Bortezomib blocks degradation of multi-proteins, including regulators of cell cycle, anti-apoptosis, and inflammation, as well as immune surveillance. In MM cells, bortezomib directly induces cell stress response followed by activation of c-Jun NH_2 terminal kinase/stress-activated protein kinase and triggers cell cycle arrest, followed by caspase-dependent apoptosis. Bortezomib also modulates activities of non-MM cellular components, including stromal cells and osteoblasts in the bone marrow milieu. Importantly, combination treatment strategies, including histone deacetylase inhibitors, Akt inhibitor, lenalidomide, heat shock protein 90 inhibitors, and aurora kinase inhibitors demonstrate significant anti-MM activities both in preclinical and clinical studies.

1 Introduction

Lysosom- and proteasome-mediated protein degradation are two major mechanisms maintaining intracellular protein catabolism, and recent studies have both defined the mechanisms of protein degradation and identified opportunities for therapeutic applications.

T. Hideshima (✉), P.G. Richardson, and K.C. Anderson
Dana-Farber Cancer Institute, Boston, MA 02115, USA
e-mail: teru_hideshima@dfci.harvard.edu, Paul_Richardson@dfci.harvard.edu, Kenneth_Anderson@dfci.harvard.edu

I.M. Ghobrial et al. (eds.), *Bortezomib in the Treatment of Multiple Myeloma*, Milestones in Drug Therapy,
DOI 10.1007/978-3-7643-8948-2_2,

Ubiquitin (Ub) is a highly conserved small protein composed of 76 amino acids found only in eukaryotic organisms. The C-terminus of ubiquitin forms an isopeptide bond with the amino group of a lysine side chain of target protein. Attachment of multiple copies of ubiquitin targets the protein for degradation by the large intracellular protease known as the 26S proteasome. The 26S proteasome is an ATP-dependent, multifunctional proteolytic complex that differs in many respects from typical proteolytic enzymes. It consists of a proteolytic core, the 20S (720 kDa) proteasome, sandwiched between two 19S (890 kDa) regulatory complexes. The 19S proteasome regulatory complexes control the access of substrates to the proteolytic core. The 20S proteasome, a multicatalytic protease, forms a hollow cylinder comprised of four stacked rings. Each outer ring is composed of seven different α-subunits and each inner ring is composed of seven distinct β-subunits. Moreover, each β-ring contains caspase-like, trypsin-like, and chymotrypsin-like proteolytically active sites. The 20S proteasome degrades oligonucleotide and protein substrates by endoproteolytic cleavage. The immunoproteasome has alternative β forms (β1i, β2i, and β5i), is expressed in hematopoietic cells (i.e., monocytes and lymphocytes) in response to exposure to pro-inflammatory stimuli (i.e., interferon-γ), and has an important role for processing peptide antigens for MHC class I presentation.

The ubiquitin–proteasome pathway (UPP) is therefore the major non-lysosomal proteolytic system in the cytosol and nucleus of all eukaryotic cells. UPP triggers degradation of proteins involved in cell cycle progression, apoptosis, transcription, inflammation, as well as immune surveillance. It also degrades mutant, damaged, and unfolded/misfolded proteins. Degradation of proteins via UPP is processed via multistep mechanisms. First, free Ub is activated in an ATP-dependent manner by Ub-activating enzyme (E1), thereby forming a complex with Ub. Second, Ub is transferred to one of many distinct Ub-conjugating enzymes (E2), which directly ubiquitinate substrate proteins. Third, E3 enzymes, which are specific to proteins and Ub–protein ligases, attach the small Ub moiety to lysine residues of acceptor proteins. The ubiquitinated proteins are then shuttled to the proteasome for degradation. Therefore, not only the proteasome but also these enzymes can be targeted as a novel therapeutic strategy.

There are different types of proteasome inhibitors classified according to the reversible or irreversible inhibition of chymotrypsin-like, trypsin-like, and/or caspase-like activities. The reversible inhibitors include peptide aldehydes (MG132) and peptide boronates (MG-262, bortezomib). The irreversible inhibitors include lactacystin and its derivatives, peptide vinyl sulfones, and peptide epoxyketones. These proteasome inhibitors block degradation of proteins, which are ubiquitinated. Bortezomib reversibly targets the chymotryptic activity. Recently, a new proteasome inhibitor NPI-0052 has been discovered during the fermentation of Salinospora sp., a marine Gram-positive actinomycete. NPI-0052 irreversibly inhibits chymotryptic, trypsin-like, and caspase-like activities [1]. Another second generation proteasome inhibitor is Carfilzomib (PR-171), a novel epoxomicin-related proteasome inhibitor, which irreversibly blocks chymotrypsin-like activity [2].

2 Biologic Sequelae of Proteasome Inhibition by Bortezomib (Velcade®)

2.1 Targeting MM Cells

As described above, the UPP is a major proteolytic system regulating a broad spectrum of proteins including cyclins and cyclin-dependent kinase (CDK) inhibitors (i.e., $p21^{Cip1}$ and $p27^{Kip1}$) as well as anti-apoptotic proteins, thereby regulating cell cycle progression and apoptosis [3, 4]. Bortezomib represents a class of peptide boronate proteasome inhibitors, which reversibly inhibits chymotryptic activity of the 26S proteasome [3, 5]. Although bortezomib is a reversible inhibitor of the proteasome, our studies show that MM cells are committed to apoptosis even after short-time exposure (2–4 h) to bortezomib [6].

The initial rationale to use bortezomib in MM is inhibition of NF-κB activity, since NF-κB plays a crucial role in the pathogenesis in cancer, including MM. The NF-κB complex is a dimer comprised of different combinations of Rel family proteins, including p65 (RelA), RelB, c-Rel, p50 (NFκB1), and p52 (NFκB2). Previous studies have revealed that NF-κB activity is mediated via two distinct, canonical and non-canonical, pathways [7–9]. We have recently shown that both canonical and non-canonical pathways are activated by co-culture with bone marrow stromal cells (BMSCs) [10]. In the canonical pathway, NF-κB is typically a heterodimer composed of p50 and p65 subunits [11], and its activity is inhibited by association with IκB family proteins [12]. Following stimulation by various factors, IκB protein is phosphorylated by IκB kinase (IKK), typically IKKβ. Phosphorylated IκB is subsequently polyubiquitinated and degraded by the 26S proteasome [13, 14], which allows p50/p65 NF-κB nuclear translocation. Bortezomib inhibits degradation of IκBα and therefore blocks NF-κB activity. However; we have recently shown that bortezomib activates canonical NF-κB pathway in MM cells both in vitro and in vivo in a xenograft model, associated with downregulation of IκBα [15]. These results strongly suggest that NF-κB may not be a major target of bortezomib in MM cells (Fig. 1).

Our in vitro studies have confirmed that bortezomib induces apoptosis is mediated by phosphorylation of c-Jun NH2-terminal kinase followed by activation of caspase-8, -9, -3, and poly (ADP-ribose) polymerase (PARP) cleavage, even in conventional drug (dexamethasone, Dex; melphalan, Mel; doxorubicin, Dox) resistant MM cell lines, as well as freshly isolated MM patient tumor cells [6, 16, 17]. Bortezomib also downregulates expression of adhesion molecules on MM cells and BMSCs and related binding, blocks constitutive and MM cell adhesion-induced cytokine secretion in BMSCs [18], and inhibits angiogenesis. Moreover, Bortezomib inhibits DNA repair by cleavage of DNA-dependent protein kinase catalytic subunit (DNA-PKcs) and ATM [17]. Importantly, treatment of MM cell lines resistant to DNA damaging agents (Mel, Dox) with those agents to which they are resistant, followed 12–24 h later with sublethal doses of Bortezomib, can inhibit repair of

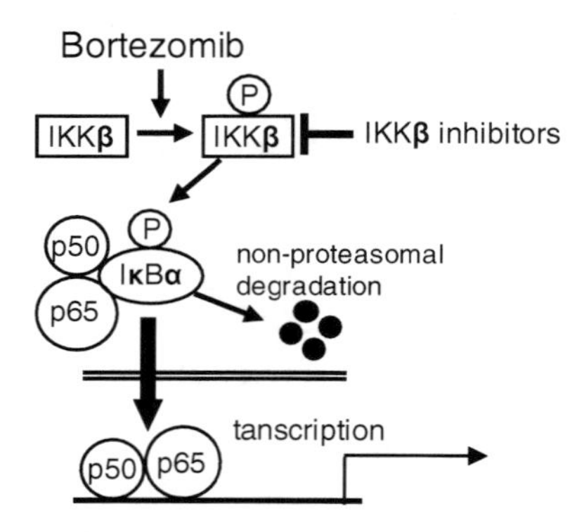

Fig. 1 Possible mechanism whereby bortezomib triggers canonical NF-κB activation. Bortezomib either directly or indirectly activates IKKβ, which subsequently phosphorylates IκBα, an inhibitor of p50/p65. After non-proteasomal degradation of IκBα, p50/p65 translocates to nucleus. IKKβ inhibitors block downregulation of IκBα and NF-κB activity, thereby enhancing bortezomib-induced cytotoxicity

DNA damage and restore drug sensitivity [19]. Bortezomib also induces caspase-dependent cleavage of gp130 (CD130), the β-subunit of IL-6 receptor, thereby abrogating IL-6–mediated downstream extracellular signal-regulated kinases (ERK), Janus kinase (JAK)2/signal transducers and activators of transcription 3 (STAT3), and phosphatidylinositol-3 kinase (PI3-K)/Akt signaling pathways, mediating growth, survival, and drug resistance, respectively, in the BM milieu [20].

Previous studies have shown that bortezomib targets the unfolded protein response (UPR) and XBP-1 in MM cells. Bortezomib suppresses the activity of IRE1α to impair the generation of spliced (active) and simultaneously stabilize the unspliced (acts as dominant negative) XBP-1 species (Fig. 2). Importantly, MM cells rendered functionally deficient in XBP-1 undergo increased apoptosis in response to ER stress [21]. Recent studies also demonstrated that bortezomib upregulates BiP, CHOP, and XBP-1, suggesting that it triggers endoplasmic reticulum (ER) stress response. Moreover, bortezomib-triggered apoptosis is dependent on caspase-2 activation, which is associated with ER stress and required for breakdown of mitochondrial transmembrane potential, release of cytochrome-c, and downstream caspase-9 activation [22]. Finally, bortezomib also significantly inhibits human MM cell growth, decreases tumor-associated angiogenesis, and prolongs host survival in human MM cells in SCID mice [23].

2.2 *Targeting the BM Microenvironment*

Bortezomib targets not only MM cells but also cellular components in the BM milieu. IL-6 is one of the most important cytokines mediating MM cell proliferation, survival, and drug resistance [24, 25]. Importantly, constitutive and MM cell adhesion-mediated IL-6 transcription and secretion from BMSCs are regulated by NF-κB [26, 27], which is inhibited by bortezomib [27]. Myeloma causes osteolytic

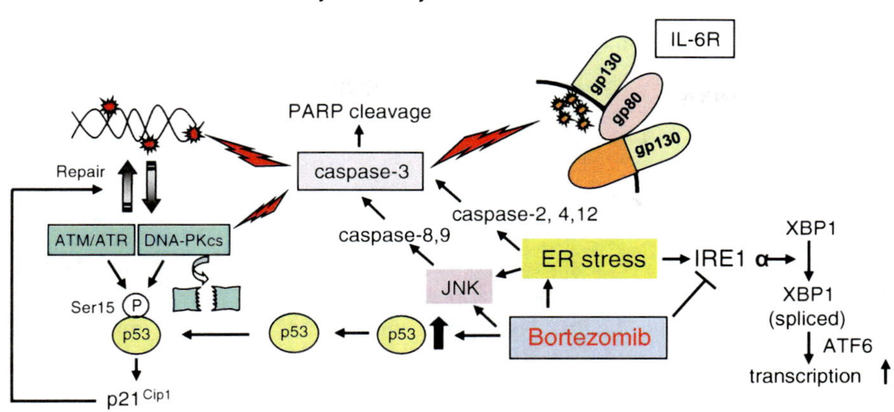

Fig. 2 Mechanism of action of bortezomib-induced cytotoxicity in MM cells. Bortezomib induces upregulation of p53 expression by inhibiting UPP. Bortezomib directly or indirectly (via ER stress) induces JNK activation, followed by activation caspases. Caspase-3 induces DNA damage, which activates p53. Activation caspase-3 also cleaves DNA-PKcs and ATM/ATR proteins, as well as gp130, resulting in impaired DNA repair and response to IL-6, respectively. Although bortezomib triggers ER stress, it also blocks IRE1α and XBP1 splicing, thereby inhibiting its transcriptional activity

bone disease and the effects of bortezomib in bone remodeling, specifically on osteoblasts and osteoclasts, have recently been reported [28, 29]. Specifically, bortezomib significantly induced a stimulatory effect on osteoblast markers in human mesenchymal stem cells (MSCs) without affecting the number of osteoblast progenitors in BM cultures or the viability of mature osteoblasts, associated with upregulated runt-related transcription factor 2 (Runx2)/Cbfa1 activity in human osteoblast progenitors and osteoblasts. Specifically, Runx2/Cbfa1-positive osteoblastic cells are observed in MM patients responding to bortezomib treatment [29]. Other studies also confirm that bortezomib induces MSCs to preferentially undergo osteoblastic differentiation by modulation of the bone-specifying transcription factor Runx-2. Mice implanted with MSCs showed increased bone formation when recipients received low doses of bortezomib [30]. Recent studies have also shown that bortezomib promotes matrix mineralization and calcium deposition by osteoprogenitor cells and primary MSCs via Wnt-independent activation of β-catenin/TCF signaling, by stabilization of β-catenin. Specifically, nuclear translocation of stabilized β-catenin was associated with β-catenin/TCF transcriptional activity that was independent of the effects of Wnt ligand receptor-induced signaling or GSK3β activation; conversely, blockade of β-catenin/TCF signaling attenuated bortezomib-induced matrix mineralization [31]. In contrast, bortezomib inhibits differentiation and bone resorption activity of osteoclasts. The mechanisms of action targeting early osteoclast differentiation was related to the inhibition of p38MAPK pathways, whereas targeting the later phase of differentiation and activation was due to inhibition of p38MAPK, AP-1, and NF-κB activation [32]. Bone marrow angiogenesis also plays an important role in the pathogenesis and progression in MM [33–35], and Bortezomib inhibits the proliferation of MM endothelial cells and angiogenesis, associated with downregulation of VEGF and Angiopoietin1/2 [36].

3 Phamacogenomics and Cytogenetics

The results of gene expression profiling in MM cell lines treated with bortezomib have already been reported. In preclinical studies, bortezomib downregulates growth/survival signaling pathways, while it upregulates molecules implicated in proapoptotic cascades and heat-shock proteins, as well as ubiquitin/proteasome pathway member proteins [37]. In clinical studies, chromosome13 deletion [del (13)] is associated with resistance to conventional treatments and poor prognosis in MM [38]. Cytogenetic studies have shown that partial or complete deletion of chromosome 13 is present in 15–20% of newly diagnosed MM patients [39]. Importantly, bortezomib overcomes the poor prognosis conferred by del(13) in the phase 2 (SUMMIT) and phase 3 (APEX) trials in the setting of relapsed and refractory MM [40]. Other studies also report that cytogenetic abnormalities including del(13), t(11;14), t(4;14), and CKS1B expression do not affect progression-free survival and overall survival in relapsed/refractory MM patients treated with bortezomib [41]. Taken together, these results suggest that bortezomib is an effective salvage therapy for refractory/replaced MM regardless of genetic risk factors.

4 Bortezomib-Based Novel Combination Strategies

4.1 Histone Deacetylase Inhibitors

Histone deacetylases (HDACs) are the enzymes that have a crucial role in the epigenetic regulation of gene expression. Recent studies have revealed that HDACs are promising therapeutic targets for cancer treatment, including MM [42]. Inhibition of histone deacetylation triggers gene transcription, and HDAC inhibitors therefore could induce transcription of both positive and negative regulators of cell proliferation/survival. As described above, intracellular protein catabolism is maintained by two different systems [43]. Specifically, polyubiquitinated proteins are degraded via both proteasomes and aggresomes in which proteins are processed by lysosomes. Importantly, HDAC6 is an essential component required for aggresome formation [44]. Therefore, inhibition of both mechanisms of protein catabolism induces accumulation of ubiquitinated proteins, followed by cell stress and cytotoxicity in MM cells [43] (Fig. 3). Indeed, we have shown that bortezomib plus

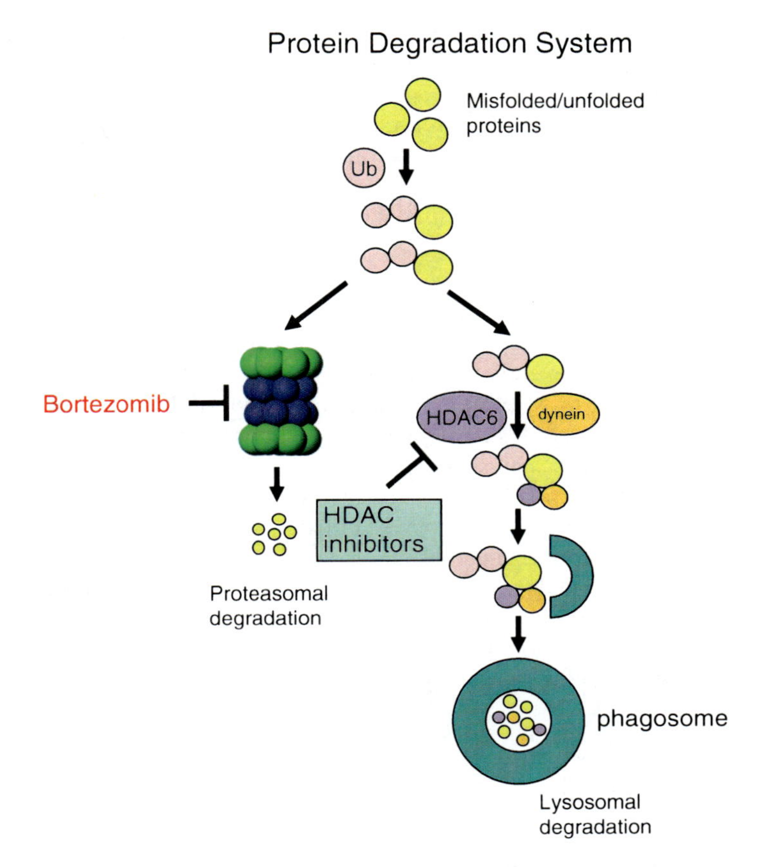

Fig. 3 Possible ubiquitinated protein catabolism in MM cells. Misfolded/unfolded proteins become polyubiquitinated and normally degraded by proteasomes. However, these proteins can escape degradation due to abnormal or pathological conditions and form toxic aggregates. These misfolded and aggregated proteins are recognized and bound by HDAC6 through the presence of polyubiquitin chains. This allows for the loading of polyubiquitinated misfolded protein cargo onto the dynein motor complex by HDAC6. The polyubiquitinated cargo-HDAC6-dynein motor complex then travels to the aggresome, where the misfolded and aggregated proteins are processed and degraded by lysosomes, clearing the cell of cytotoxic protein aggregates. HDAC inhibitors can block HDAC6 function, thereby impairing aggresome formation. Dual blockade of proteasome pathway by bortezomib and aggresome pathway by HDAC inhibitors significantly induces accumulation of polyubiquitinated proteins and triggers synergistic cytotoxicity in MM cells

HDAC6-specific inhibitor tubacin [45] or pan-HDAC inhibitor LBH589 induces significant JNK activation followed by MM cell death mediated via caspase-dependent apoptosis, without toxicity in normal peripheral blood mononuclear cells (PBMCs) [46, 47]. The synergistic anti-MM activities of bortezomib with other histone deacetylase inhibitors have also been reported [48–50].

Most recently, efficacy of this combination treatment against MM-related bone disease has been reported. In this study, hydroxamate-based histone deacetylase inhibitor JNJ-26481585 with bortezomib showed more pronounced reduction of osteoclasts and increase of osteoblasts, trabecular bone volume, and trabecular number than bortezomib as a single agent [51]. These results indicate that bortezomib combined with histone deacetylase inhibitors targets not only MM cells but also other cellular components in the BM microenvironment.

4.2 Perifosine

Perifosine is a synthetic alkylphospholipid, a new class of anti-tumor agents, that targets cell membranes, potently inhibits Akt activation, and triggers apoptosis even of MM cells adherent to BMSCs [52]. Although bortezomib alone has anti-MM activities, we found that perifosine augments its cytotoxicity in MM cells. There are two distinct possible molecular mechanisms inducing potent cytotoxicity associated with upregulated JNK activation and cleavage of caspase-8/PARP. First, bortezomib induces activation of Akt, which plays a crucial role in MM cells drug resistance [53, 54]; conversely, perifosine completely blocks bortezomib-induced Akt activation. Second, perifosine induces ERK activation, which promotes MM cell proliferation [55], and bortezomib completely abrogates perifosine-induced ERK activation [52] (Fig. 4). Taken together, this combination potently inhibits both ERK and Akt-signaling cascades, resulting in synergistic cytotoxicity in MM cells.

4.3 Hsp90 Inhibitors

Hsp90 is a molecular chaperone that interacts with client intracellular proteins to facilitate intracellular trafficking, conformational maturation, and 3-dimensional folding required for protein function. We and others have shown that geldanamycin and its analogs (i.e., 17AAG), as well as other inhibitor hsp 90 inhibitors show anti-MM activities as single agents, which is markedly augmented in combination with bortezomib [56–59]. However, molecular mechanisms, whereby bortezomib with Hsp90 inhibitors induce strong cytotoxicity against MM cells, have not been fully elucidated. Bortezomib upregulates Hsp90 and its client proteins (i.e., Akt, survivin), mediating MM cell survival, and anti-apoptosis; conversely, Hsp90 inhibitors downregulate function and/or expression of these proteins. Hsp90 inhibitors also decrease growth factor receptors (IGF-1, IL-6) which could reduce bortezomib-induced cytotoxicity [57]. Interestingly, other studies have also shown that Hsp90 inhibitor plus bortezomib combination simultaneously disrupts hsp90 and proteasome function, promotes the accumulation of aggregated, ubiquitinated proteins, and results in enhanced anti-tumor activity in breast cancer cells [60]. Previous studies have also shown that Hsp90 inhibitor induces UPR [61]. Since UPR enhances bortezomib-induced cytotoxicity, this may be another mechanism, whereby Hsp90 inhibitor augments bortezomib-induced cytotoxicity.

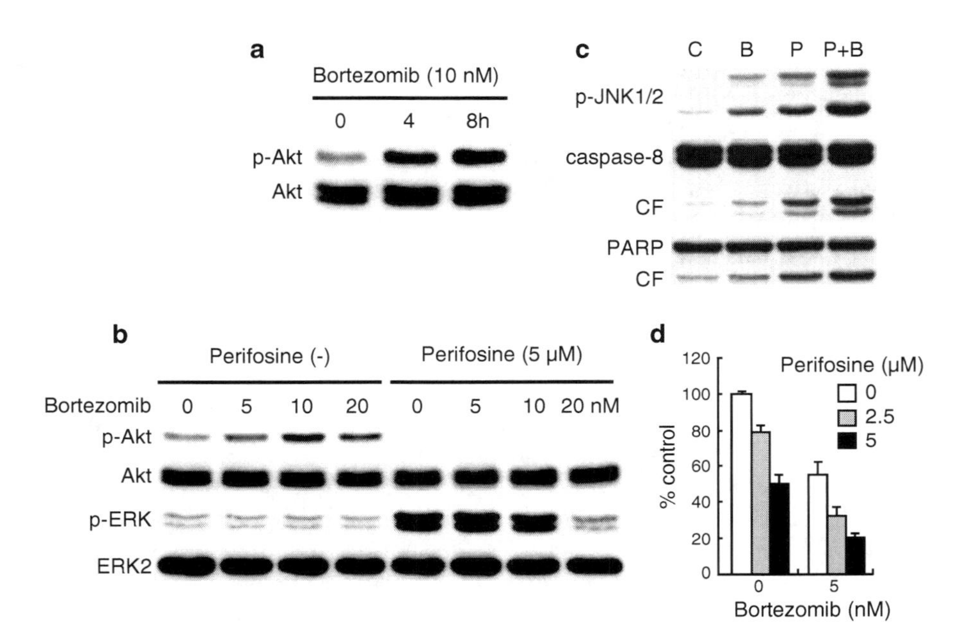

Fig. 4 Perifosine enhances cytotoxicity of bortezomib. (**a**) MM cells were cultured with bortezomib (10 nM) for 4 and 8 h. Whole-cell lysates were subjected to Western blotting with p-Akt and Akt Abs. (**b**) MM cells were cultured with bortezomib (5, 10, 20 nM) in the presence (5 μM) or absence of perifosine. Whole-cell lysates were subjected to Western blotting with p-Akt, Akt, p-ERK, and ERK2 Abs. (**c**) MM cells were cultured for 8 h with control media, perifosine (5 μM), bortezomib (10 nM), or perifosine (5 μM) plus bortezomib (10 nM). Cells were then lysed and subjected to Western blotting using anti-p-JNK1/2, caspase-8, and PARP Abs. (**d**) MM cells were cultured for 24 h with control media (*open square*); and with 2.5 μM (*gray filled square*) or 5 μM (*black filled square*) perifosine; in the presence or absence of bortezomib (5 and 7.5 nM). Cytotoxicity was assessed by MTT assay; data represent mean (±SD) of quadruplicate cultures

4.4 Lenalidomide (Revlimid®) Plus Dexamethasone

Lenalidomide is an immunomudulatory derivative of thalidomide and its anti-MM activities have been demonstrated in preclinical [62–65] and clinical studies [66–68]. Moreover, lenalidomide plus dexamethasone also has significant anti-MM activity in vitro [62] and in patients [69, 70]. Three combination of bortezomib with lenalidomide plus dexamethasone has been evaluated in clinical trials demonstrating remarkable clinical efficacy [71, 72]. Although molecular mechanisms of action, thereby inducing growth inhibition by this combination, are not totally understood, different apoptotic signaling pathways (intrinsic vs. extrinsic) induced by these agents, at least in part, account synergistic anti-MM activities.

4.5 *IKKβ Inhibitors*

As described above, bortezomib induces phosphorylation and downregulation of IκBα expression, thereby activating canonical NF-κB pathway. Therefore, inhibition of canonical NF-κB pathway could enhance bortezomib-induced cytotoxicity. We have shown that IKKβ inhibitors block IKKβ activation and downregulation of IκBα protein expression induced by bortezomib, resulting in inhibition of canonical NF-κB activation. Importantly, IKKβ inhibitors further enhance bortezomib-induced cytotoxicity in MM cells [15].

4.6 *Aurora Kinase Inhibitors*

Aurora kinases (A and B) are serine/threonine kinases playing crucial role in mitosis and cell proliferation. Recent studies have shown that specific inhibitors of aurora kinases inhibit MM cell growth in vitro and in vivo mouse models [73–76]. In MM patients, aurora kinases are overexpressed who have poor prognosis with high centrosome index (CI) [74]. Since MM cell lines with high CI index are more sensitive to treatment with aurora kinase inhibitor [74], and ectopic overexpression of aurora kinase does not affect bortezomib [73], combination of bortezomib with aurora kinase inhibitors may represent a novel therapeutic strategy in patients with high CI.

5 Natural Products that Inhibit Bortezomib

Some components of diet and/or dietary supplements may affect efficiency of bortezomib. Since previous studies have shown that antioxidant (i.e., LNAC) inhibits bortezomib-induced cytotoxicity [77], some of antioxidants in diet and/or dietary supplements may affect anti-tumor activity of bortezomib. Specifically, recent studies have shown that daily oral intake of vitamin C (ascorbic acid) blocks anti-MM activities of bortezomib, associated with 20S proteasome activity [78]. Moreover, plasma collected from healthy volunteers taking 1 g/day vitamin C reduced bortezomib-induced MM cell death in vitro. This antagonistic effect of vitamin C against proteasome inhibitors is specific for boronate class of inhibitors. These results suggest that patients receiving treatment with bortezomib should avoid taking vitamin C dietary supplements [79]. Other studies have also shown that green tea constituents, in particular epigallocatechin gallate (EGCG) and other polyphenols with 1,2-benzenediol moieties, prevented tumor cell death induced by boronate class proteasome inhibitors, including bortezomib, in vitro and in vivo [80].

Reference

1. Chauhan D, Catley L, Li G et al (2005) A novel orally active proteasome inhibitor induces apoptosis in multiple myeloma cells with mechanisms distinct from Bortezomib. Cancer Cell 8:407–419
2. Kuhn DJ, Chen Q, Voorhees PM et al (2007) Potent activity of carfilzomib, a novel, irreversible inhibitor of the ubiquitin-proteasome pathway, against preclinical models of multiple myeloma. Blood 110:3281–3290
3. Kisselv AF, Goldberg AL (2001) Proteasome inhibitors: from research tools to drug candidates. Chem Biol 21:1–20
4. Hideshima T, Richardson PG, Anderson KC (2003) Targeting proteasome inhibition in hematologic malignancies. Rev Clin Exp Hematol 7:191–204
5. Adams J (2004) The proteasome: a suitable antineoplastic target. Nat Rev Cancer 4:349–360
6. Hideshima T, Richardson P, Chauhan D et al (2001) The proteasome inhibitor PS-341 inhibits growth, induces apoptosis, and overcomes drug resistance in human multiple myeloma cells. Cancer Res 61:3071–3076
7. Jost PJ, Ruland J (2007) Aberrant NF-kappaB signaling in lymphoma: mechanisms, consequences, and therapeutic implications. Blood 109:2700–2707
8. Keats JJ, Fonseca R, Chesi M et al (2007) Promiscuous mutations activate the noncanonical NF-kappaB pathway in multiple myeloma. Cancer Cell 12:131–144
9. Annunziata CM, Davis RE, Demchenko Y et al (2007) Frequent engagement of the classical and alternative NF-kappaB pathways by diverse genetic abnormalities in multiple myeloma. Cancer Cell 12:115–130
10. Hideshima T, Chauhan D, Kiziltepe T et al (2009) Biologic sequelae of I{kappa}B kinase (IKK) inhibition in multiple myeloma: therapeutic implications. Blood 113:5228–5236
11. Baldwin AS Jr (1996) The NF-kB and I kB proteins: new discoveries and insights. Annu Rev Immunol 14:649–683
12. Beg AA, Baldwin AS Jr (1993) The IkB proteins: multifunctional regulators of Rel NF-kB transcription factors. Genes Dev 7:2064–2070
13. Zandi E, Chen Y, Karin M (1998) Direct phosphorylation of IkappaB by IKKalpha and IKKbeta: discrimination between free and NF-kappaB-bound substrate. Science 281: 1360–1363
14. DiDonato JA, Hayakawa M, Rothwarf DM, Zandi E, Karin M (1997) A cytokine-responsive IkB kinase that activates the transcription factor NF-kB. Nature 388:548–554
15. Hideshima T, Ikeda H, Chauhan D et al (2009) Bortezomib induces canonical nuclear factor-kappaB activation in multiple myeloma cells. Blood 114:1046–1052
16. Akiyama M, Hideshima T, Hayashi T et al (2002) Cytokines modulate telomerase activity in a human multiple myeloma cell line. Cancer Res 62:3876–3882
17. Hideshima T, Mitsiades C, Akiyama M et al (2003) Molecular mechanisms mediating antimyeloma activity of proteasome inhibitor PS-341. Blood 101:1530–1534
18. Hideshima T, Chauhan D, Schlossman RL, Richardson PR, Anderson KC (2001) Role of TNF-a in the pathophysiology of human multiple myeloma: therapeutic applications. Oncogene 20:4519–4527
19. Mitsiades N, Mitsiades CS, Richardson PG et al (2003) The proteasome inhibitor PS-341 potentiates sensitivity of multiple myeloma cells to conventional chemotherapeutic agents: therapeutic applications. Blood 101:2377–2380
20. Hideshima T, Chauhan D, Hayashi T et al (2003) Proteasome Inhibitor PS-341 abrogates IL-6 triggered signaling cascades via caspase-dependent downregulation of gp130 in multiple myeloma. Oncogene 22:8386–8393
21. Lee AH, Iwakoshi NN, Anderson KC, Glimcher LH (2003) Proteasome inhibitors disrupt the unfolded protein response in myeloma cells. Proc Natl Acad Sci USA 100:9946–9951

22. Gu H, Chen X, Gao G, Dong H (2008) Caspase-2 functions upstream of mitochondria in endoplasmic reticulum stress-induced apoptosis by bortezomib in human myeloma cells. Mol Cancer Ther 7:2298–2307
23. LeBlanc R, Catley LP, Hideshima T et al (2002) Proteasome inhibitor PS-341 inhibits human myeloma cell growth in vivo and prolongs survival in a murine model. Cancer Res 62:4996–5000
24. Klein B, Zhang XG, Lu XY, Bataille R (1995) Interleukin-6 in human multiple myeloma. Blood 85:863–872
25. Hideshima T, Podar K, Chauhan D, Anderson KC (2005) Cytokines and signal transduction. Best Pract Res Clin Haematol 18:509–524
26. Chauhan D, Uchiyama H, Akbarali Y et al (1996) Multiple myeloma cell adhesion-induced interleukin-6 expression in bone marrow stromal cells involves activation of NF-kB. Blood 87:1104–1112
27. Hideshima T, Chauhan D, Richardson P et al (2002) NF-kB as a therapeutic target in multiple myeloma. J Biol Chem 277:16639–16647
28. Heider U, Kaiser M, Muller C et al (2006) Bortezomib increases osteoblast activity in myeloma patients irrespective of response to treatment. Eur J Haematol 77:233–238
29. Giuliani N, Morandi F, Tagliaferri S et al (2007) The proteasome inhibitor bortezomib affects osteoblast differentiation in vitro and in vivo in multiple myeloma patients. Blood 110: 334–338
30. Mukherjee S, Raje N, Schoonmaker JA et al (2008) Pharmacologic targeting of a stem/ progenitor population in vivo is associated with enhanced bone regeneration in mice. J Clin Invest 118:491–504
31. Qiang YW, Hu B, Chen Y et al (2009) Bortezomib induces osteoblast differentiation via Wnt-independent activation of beta-catenin/TCF signaling. Blood 113:4319–4330
32. von Metzler I, Krebbel H, Hecht M et al (2007) Bortezomib inhibits human osteoclastogenesis. Leukemia 21:2025–2034
33. Vacca A, Ribatti D, Roncali L et al (1994) Bone marrow angiogenesis and progression in multiple myeloma. Br J Haematol 87:503–508
34. Vacca A, Ribatti D, Presta M et al (1999) Bone marrow neovascularization, plasma cell angiogenic potential, and matrix metalloproteinase-2 secretion parallel progression of human multiple myeloma. Blood 93:3064–3073
35. Kumar S, Fonseca R, Dispenzieri A et al (2002) Bone marrow angiogenesis in multiple myeloma: effect of therapy. Br J Haematol 119:665–671
36. Roccaro AM, Hideshima T, Raje N et al (2006) Bortezomib mediates antiangiogenesis in multiple myeloma via direct and indirect effects on endothelial cells. Cancer Res 66:184–191
37. Mitsiades N, Mitsiades CS, Poulaki V et al (2002) Molecular sequelae of proteasome inhibition in human multiple myeloma cells. Proc Natl Acad Sci USA 99:14374–14379
38. Tricot G, Barlogie B, Jagannath S et al (1995) Poor prognosis in multiple myeloma is associated only with partial or complete deletions of chromosome 13 or abnormalities involving 11q and not with other karyotype abnormalities. Blood 86:4250–4256
39. Shaughnessy J, Tian E, Sawyer J et al (2000) High incidence of chromosome 13 deletion in multiple myeloma detected by multiprobe interphase FISH. Blood 96:1505–1511
40. Jagannath S, Richardson PG, Sonneveld P et al (2007) Bortezomib appears to overcome the poor prognosis conferred by chromosome 13 deletion in phase 2 and 3 trials. Leukemia 21:151–157
41. Chang H, Trieu Y, Qi X, Xu W, Stewart KA, Reece D (2007) Bortezomib therapy response is independent of cytogenetic abnormalities in relapsed/refractory multiple myeloma. Leuk Res 31:779–782
42. Mitsiades N, Mitsiades CS, Richardson PG et al (2003) Molecular sequelae of histone deacetylase inhibition in human malignant B cells. Blood 101:4055–4062
43. Hideshima T, Bradner JE, Chauhan D, Anderson KC (2005) Intracellular protein degradation and its therapeutic implications. Clin Cancer Res 11:8530–8533

44. Kawaguchi Y, Kovacs JJ, McLaurin A, Vance JM, Ito A, Yao TP (2003) The deacetylase HDAC6 regulates aggresome formation and cell viability in response to misfolded protein stress. Cell 115:727–738
45. Haggarty SJ, Koeller KM, Wong JC, Grozinger CM, Schreiber SL (2003) Domain-selective small-molecule inhibitor of histone deacetylase 6 (HDAC6)-mediated tubulin deacetylation. Proc Natl Acad Sci USA 100:4389–4394
46. Hideshima H, Bradner JE, Wong J et al (2005) Small molecule inhibition of proteasome and aggresome function induces synergistic anti-tumor activity in multiple myeloma. Proc Natl Acad Sci USA 102:8567–8572
47. Catley L, Weisberg E, Kiziltepe T et al (2006) Aggresome induction by proteasome inhibitor bortezomib and alpha-tubulin hyperacetylation by tubulin deacetylase (TDAC) inhibitor LBH589 are synergistic in myeloma cells. Blood 108:3441–3449
48. Mitsiades CS, Mitsiades NS, McMullan CJ et al (2004) Transcriptional signature of histone deacetylase inhibition in multiple myeloma: biological and clinical implications. Proc Natl Acad Sci USA 101:540–545
49. Pei XY, Dai Y, Grant S (2004) Synergistic induction of oxidative injury and apoptosis in human multiple myeloma cells by the proteasome inhibitor bortezomib and histone deacetylase inhibitors. Clin Cancer Res 10:3839–3852
50. Feng R, Oton A, Mapara MY, Anderson G, Belani C, Lentzsch S (2007) The histone deacetylase inhibitor, PXD101, potentiates bortezomib-induced anti-multiple myeloma effect by induction of oxidative stress and DNA damage. Br J Haematol 139:385–397
51. Deleu S, Lemaire M, Arts J et al (2009) Bortezomib alone or in combination with the histone deacetylase inhibitor JNJ-26481585: effect on myeloma bone disease in the 5T2MM murine model of myeloma. Cancer Res 69:5307–5311
52. Hideshima T, Catley L, Yasui H et al (2006) Perifosine, an oral bioactive novel alkylphospholipid, inhibits Akt and induces in vitro and in vivo cytotoxicity in human multiple myeloma cells. Blood 107:4053–4062
53. Hideshima T, Nakamura N, Chauhan D, Anderson KC (2001) Biologic sequelae of interleukin-6 induced PI3-K/Akt signaling in multiple myeloma. Oncogene 20:5991–6000
54. Hideshima T, Catley L, Raje N et al (2007) Inhibition of Akt induces significant downregulation of survivin and cytotoxicity in human multiple myeloma cells. Br J Haematol 138:783–791
55. Ogata A, Chauhan D, Urashima M, Teoh G, Treon SP, Anderson KC (1997) Blockade of mitogen-activated protein kinase cascade signaling in interleukin-6 independent multiple myeloma cells. Clin Cancer Res 3:1017–1022
56. Mitsiades CS, Mitsiades N, Poulaki V, Akiyama M, Treon SP, Anderson KC (2001) The HSP90 molecular chaperone as a novel therapeutic target in hematologic malignancies. Blood 98:377a
57. Mitsiades CS, Mitsiades NS, McMullan CJ et al (2006) Antimyeloma activity of heat shock protein-90 inhibition. Blood 107:1092–1100
58. Stuhmer T, Zollinger A, Siegmund D et al (2008) Signalling profile and antitumour activity of the novel Hsp90 inhibitor NVP-AUY922 in multiple myeloma. Leukemia 22:1604–1612
59. Okawa Y, Hideshima T, Steed P et al (2009) SNX-2112, a selective Hsp90 inhibitor, potently inhibits tumor cell growth, angiogenesis, and osteoclastogenesis in multiple myeloma and other hematological tumors by abrogating signaling via Akt and ERK. Blood 113(4):846–855
60. Mimnaugh EG, Xu W, Vos M et al (2004) Simultaneous inhibition of hsp 90 and the proteasome promotes protein ubiquitination, causes endoplasmic reticulum-derived cytosolic vacuolization, and enhances antitumor activity. Mol Cancer Ther 3:551–566
61. Davenport EL, Moore HE, Dunlop AS et al (2007) Heat shock protein inhibition is associated with activation of the unfolded protein response (UPR) pathway in myeloma plasma cells. Blood 110(7):2641–2649
62. Hideshima T, Chauhan D, Shima Y et al (2000) Thalidomide and its analogues overcome drug resistance of human multiple myeloma cells to conventional therapy. Blood 96:2943–2950

63. Davies FE, Raje N, Hideshima T et al (2001) Thalidomide and immunomodulatory derivatives augment natural killer cell cytotoxicity in multiple myeloma. Blood 98:210–216
64. Mitsiades N, Mitsiades CS, Poulaki V et al (2002) Apoptotic signaling induced by immunomodulatory thalidomide analogs in human multiple myeloma cells: therapeutic implications. Blood 99:4525–4530
65. Hayashi T, Hideshima T, Akiyama M et al (2005) Molecular mechanisms whereby immunomodulatory drugs activate natural killer cells: clinical application. Br J Haematol 128: 192–203
66. Richardson PG, Schlossman RL, Weller E et al (2002) Immunomodulatory drug CC-5013 overcomes drug resistance and is well tolerated in patients with relapsed multiple myeloma. Blood 100:3063–3067
67. Richardson PG, Blood E, Mitsiades CS et al (2006) A randomized phase 2 study of lenalidomide therapy for patients with relapsed or relapsed and refractory multiple myeloma. Blood 108:3458–3464
68. Richardson P, Jagannath S, Hussein M et al (2009) Safety and efficacy of single-agent lenalidomide in patients with relapsed and refractory multiple myeloma. Blood 114:772–778
69. Dimopoulos M, Spencer A, Attal M et al (2007) Lenalidomide plus dexamethasone for relapsed or refractory multiple myeloma. N Engl J Med 357:2123–2132
70. Weber DM, Chen C, Niesvizky R et al (2007) Lenalidomide plus dexamethasone for relapsed multiple myeloma in North America. N Engl J Med 357:2133–2142
71. Richardson P, Jagannath S, Jakubowiak A et al (2008) Lenalidomide, bortezomib, and dexamethasone in patients with relapsed or relapsed/refractory multiple myeloma (MM): encouraging response rates and tolerability with correlation of outcome and adverse cytogenetics in a phase II study. Blood 112:614
72. Richardson P, Jagannath S, Jakubowiak A et al (2008) Lenalidomide, bortezomib, and dexamethasone in patients with newly diagnosed multiple myeloma: encouraging efficacy in high risk groups with updated results of a phase I/II study. Blood 112:41
73. Shi Y, Reiman T, Li W et al (2007) Targeting aurora kinases as therapy in multiple myeloma. Blood 109:3915–3921
74. Chng WJ, Braggio E, Mulligan G et al (2008) The centrosome index is a powerful prognostic marker in myeloma and identifies a cohort of patients that might benefit from aurora kinase inhibition. Blood 111:1603–1609
75. Hose D, Reme T, Meissner T et al (2009) Inhibition of aurora kinases for tailored risk-adapted treatment of multiple myeloma. Blood 113:4331–4340
76. Dutta-Simmons J, Zhang Y, Gorgun G et al (2009) Aurora kinase A is a target of Wnt/{beta}-catenin involved in multiple myeloma disease progression. Blood 114(3):2699–2708
77. Yu C, Rahmani M, Dent P, Grant S (2004) The hierarchical relationship between MAPK signaling and ROS generation in human leukemia cells undergoing apoptosis in response to the proteasome inhibitor bortezomib. Exp Cell Res 295:555–566
78. Zou W, Yue P, Lin N et al (2006) Vitamin C inactivates the proteasome inhibitor PS-341 in human cancer cells. Clin Cancer Res 12:273–280
79. Perrone G, Hideshima T, Ikeda H et al (2009) Ascorbic acid inhibits antitumor activity of bortezomib in vivo. Leukemia 23(9):1679–1686
80. Golden EB, Lam PY, Kardosh A et al (2009) Green tea polyphenols block the anticancer effects of bortezomib and other boronic acid-based proteasome inhibitors. Blood 113: 5927–5937

Bortezomib and Osteoclasts and Osteoblasts

Michal T. Krauze and G. David Roodman

Abstract Bortezomib is the first-in-class proteasome antagonist approved for treatment of myeloma. It is active in newly diagnosed, relapsed, and refractory patients and is now being used as a platform for combinations with other new agents for myeloma. In addition to its anti-myeloma effect, bortezomib also targets the bone microenvironment and can inhibit osteoclast formation and stimulate osteoblast activity in patients with myeloma. Potentially, combination of bortezomib with other agents that stimulate bone formation or block bone resorption will further enhance the anti-myeloma effects of bortezomib and overcome the contribution of the tumor microenvironment to myeloma growth. In this chapter, we discuss the potential mechanisms responsible for bortezomib's effects on osteoclast and osteoblast activity in myeloma.

1 Introduction

Multiple Myeloma (MM) is a primary plasma cell malignancy that is the most frequent malignancy to involve the bone. MM is characterized by the development of a progressive and destructive osteolytic bone disease, associated with severe bone pain, pathological fractures, osteoporosis, hypercalcemia, and spinal cord compression. Over 80% of patients with MM have bone involvement during the course of their disease [1]. Patients with MM have the highest incidence of fractures

M.T. Krauze
Department of Medicine/Hematology-Oncology, University of Pittsburgh, Pittsburgh, PA, USA

G.D. Roodman (✉)
Department of Medicine/Hematology-Oncology, University of Pittsburgh, Pittsburgh, PA, USA
Veterans Affairs Pittsburgh Healthcare System, Research and Development, Department of Medicine/Hematology-Oncology, University of Pittsburgh School of Medicine, 151-U University Drive C, Rm. 2E113, Pittsburgh, PA 15240, USA
e-mail: Roodmangd@upmc.edu

I.M. Ghobrial et al. (eds.), *Bortezomib in the Treatment of Multiple Myeloma*, Milestones in Drug Therapy,
DOI 10.1007/978-3-7643-8948-2_3,

(43%) compared to breast cancer, prostate cancer, and lung cancer patients, respectively. MM patients who experienced pathologic fractures had at least a 20% increased risk of death compared to MM patients without pathologic fractures [2]. Despite many significant advances in the understanding of the biology of MM, it remains an incurable malignancy in the overwhelming majority of patients, and the destructive osteolytic bone disease is a major cause of morbidity in patients with MM.

Myeloma cells have a predisposition to group around sites of active bone resorption, and the interactions among myeloma cells, osteoclasts, stromal cells, myeloma-associated fibroblasts, and osteoblasts are crucial both for the development of the osteolytic bone disease and for myeloma cell growth and survival in the bone marrow [3, 4]. Multiple myeloma patients lose bone much more rapidly than age-matched controls. One study reported that patients with MM who received glucocorticoid-containing treatment lost approximately 6% of their bone mineral density at the lumbar spine and almost 10% of their bone mineral density at the femoral neck over a 12-month period. In contrast, age-matched controls, who were not on steroid therapy, lost essentially no bone at the lumbar spine and approximately 1% of their bone mineral density at the femoral neck over the same time period [5]. Further, the frequency of skeletal-related events is very high in patients with MM, with more than 30% of the patients having a fracture or requiring radiation therapy for bone pain on the placebo arm of a 21-month clinical trial [6]. These data clearly show that osteoclast activity is markedly increased in MM. Further, histologic studies of bone biopsies from patients with MM show that increased osteoclast activity occurs adjacent to MM cells.

One of the new therapies that have made a major impact on the survival and quality of life in MM patients is the introduction of bortezomib, which is the first-in class proteasome antagonist to come to the clinic. Bortezomib is of particular interest in MM therapy as it has shown to inhibit osteoclast resorptive activity in vitro and in vivo [7, 8]. The CREST trial examined the efficacy of 1.0 vs. 1.3 mg/m^2 of bortezomib in patients with relapsing MM [9]. The recently updated survival results of this trial showed that 1- and 5-year survival rates were 82% and 32% for the 1.0 mg/m^2 dose and 81% and 45% for the 1.3 mg/m^2 dose, respectively [10]. These results clearly demonstrated the clinical efficacy of bortezomib in MM. This chapter will discuss the clinical importance of osteoclast and osteoblast activity in MM, the effects of bortezomib in controlling bone disease in MM.

2 Role of the Bone Marrow Microenvironment in Myeloma

Normal bone constantly undergoes remodeling, which involves the resorption of bone by osteoclasts and the deposition of new bone by osteoblasts at sites of previous resorption. Osteoclasts arise from precursor cells in the monocyte/macrophage lineage that differentiate into inactive osteoclasts. The inactive osteoclasts are then activated, resorb bone, and subsequently undergo apoptosis. Both locally produced cytokines and systemic hormones normally regulate osteoclast formation and activity. In myeloma, many different components interact to increase the bone-resorptive process. These include myeloma cells themselves, bone-marrow stromal cells, and T cells present in the marrow microenvironment. Furthermore, growth factors released by the bone-resorptive process also increase the growth of myeloma cells [11]. This creates a vicious cycle, with the bone resorptive process increasing myeloma cell tumor burden, which then results in increased bone resorption.

In addition to the increased osteoclast activity, osteoblast activity is markedly suppressed in MM. Histomorphometric studies have shown that bone remodeling is uncoupled in MM with increased bone resorption and decreased or absent bone formation. Further, osteoblast apoptosis is markedly increased due to high cytokine levels and physical interaction between osteoblasts and MM cells [12]. Thus, MM patients have low levels of bone formation markers, such as alkaline phosphatase and osteocalcin [13]. This explains why bone scans underestimate the extent of MM bone disease, as bone scans reflect new bone formation.

The marrow microenvironment plays a critical role in both tumor growth and the bone destructive process in MM. MM cells home to the bone marrow by means of SDF-1 expressed by stromal cells in the marrow microenvironment and CXCR4 and CXCR7 on MM cells [14]. MM cells then bind to marrow stromal cells through adhesive interactions that activate multiple signaling pathways within the MM cells and in marrow stromal cells [15] that result in enhanced tumor growth and bone destruction. The marrow microenvironment supports the growth of MM cells and enhances osteoclastic bone destruction through expression of growth factors such as VEGF, IGF-1, IL-6, TNF-α, and RANK ligand [16]. In addition, MM cells produce IL-3 and MIP-1α, which are also potent stimulators of osteoclast formation. MM cells also block osteoblast differentiation through production of the Wnt signaling antagonists, DKK1 and sFRP2, and the cytokines IL-7 and IL-3 [17].

The RANK/RANKL signaling pathway is a critical component of the bone-remodeling process and RANK plays a role as transmembrane signaling receptor that is a member of the tumor necrosis factor (TNF) receptor superfamily. It is expressed on the surface of osteoclast precursors [18]. RANK ligand (RANKL) is expressed as a membrane-bound protein on marrow stromal cells and osteoblasts, and secreted by activated lymphocytes. Its expression is augmented by cytokines that stimulate bone resorption, such as parathyroid hormone (PTH), 1,25-$(OH)_2$ vitamin D3, and prostaglandins [19, 20]. RANKL binds to RANK receptor on osteoclast precursors and induces osteoclast formation. Rank signals through the NF-κB and JunN terminal kinase pathways and induces increased osteoclastic bone resorption and enhanced osteoclast survival [21]. The important role of RANKL in normal osteoclastogenesis has been clearly demonstrated in RANKL or RANK gene knockout mice. These animals lack osteoclasts and as a result develop severe osteopetrosis [22, 23]. All these factors play an important role in the increased growth of tumor and bone destruction in myeloma. Thus, agents that target both MM cells and the marrow microenvironment should have a greater impact on the disease than agents that only target MM cells themselves. When MM cells home to the marrow, they bind marrow stromal cells through integrins, cadherin, selectins, and syndecans expressed on the surface of the MM cells or bone [24]. Recently, MM cells have been shown to also express the surface glycoprotein, CS-1, a member of the immunoglobulin gene superfamily [25], as well as HLA-1 and β2-microglobulin that also play a role in tumor progression [26]. In particular, VLA-4 and VLA-5 bind to VCAM-1 and other receptors on marrow stromal cells to enhance signaling within the MM cells and marrow stromal cells that result in increased growth of tumor and osteoclastogenesis [27, 28].

In addition, when MM cells bind to marrow stromal cells, they become more chemoresistant [29]. These results suggest that both MM cells and the marrow microenvironment must both be targeted to eradicate MM.

3 Bortezomib in the Treatment of Bone Disease in MM

As mentioned previously, bortezomib is the first-in-class proteasome antagonist approved for treatment of myeloma. The observation of high proteasome levels in human leukemia cells and the increase of proteasome concentration in normal human mononuclear cells during blastogenic transformation induced by phytohemagglutinin, their oscillation during the cell cycle, and parallel increase with induction of DNA synthesis led to the hypothesis that proteasomes may be involved in transformation and proliferation of cells [30]. Bortezomib, formerly known as PS-341, is a boron containing molecule that specifically and reversibly inhibits the threonine residue of the 26S proteasome, an enzyme complex that plays a key role in the cell by regulating protein degradation in a controlled fashion. Proteins that are no longer required, including those involved in cell cycle control, apoptosis, and cell signaling, are tagged with ubiquitin, which directs them to the proteasome that subsequently degrades them. This process maintains the balance of inhibitory and stimulatory proteins involved in the cell cycle. Thus, inhibition of the proteasome results in a loss of the tight control of the cell cycle with a build-up of cell cycle and regulatory proteins leading to cell death [31, 32]. Other reports also suggest that bortezomib may dysregulate intracellular calcium metabolism resulting in caspase activation and apoptosis [33]. One central mechanism by which bortezomib functions in myeloma is via the inhibition of the breakdown of inhibitory kappa B (IκB) and consequently stabilization of the nuclear factor kappa B (NF-κB) complex. This prevents NF-κB translocation to the nucleus with subsequent inactivation of multiple downstream pathways known to be important in myeloma cell signaling [34]. It also decreases the adhesion of myeloma cells to stromal cells, thereby increasing MM cell sensitivity to apoptosis, as well as interrupting prosurvival paracrine and autocrine cytokine loops in the bone marrow microenvironment mediated by interleukin-6 (IL-6), insulin-like growth factor 1 (IGF-1), vascular endothelial growth factor (VEGF), and tumor necrosis factor-α (TNF-α) [35, 36]. Other effects of bortezomib in myeloma include inhibition of angiogenesis, inhibition of DNA repair, and impairment of osteoclast activity [37]. The NF-κB signaling pathway plays an important role in osteoclast formation and activity in response to RANKL. This inhibition of NF-κB is thought to be the basis for the osteoclast inhibition by bortezomib [38]. Giuliani et al. and Terpos et al. both report decrease in bone resorption markers in patients receiving bortezomib, at least in patients whose MM responded to therapy. These data suggest that bortezomib may suppress osteoclast activity in vitro as well as in vivo, although this has not been confirmed by others [39]. In preclinical models of myeloma, Pennisi et al. [40] have shown that bortezomib inhibits osteoclastogenesis. However, this has not been a universal finding and it remains to be seen if bortezomib has effects on osteoclastogenesis in vivo in patients whose MM does not respond to bortezomib. In addition to decreasing osteoclast activity in MM, bortezomib was also found to induce osteoblast differentiation and thereby stimulate new bone formation [41, 42]. The theory that bortezomib exerts its main antitumor effect in myeloma through inhibition of NF-κB has been challenged in recent years [43]. A cautionary note was introduced by work performed several years ago, showing that NF-κB inhibition did not fully account for bortezomib's cytotoxic effects in MM cells [44]. As an alternative hypothesis, Hideshima et al. proposed that cell death results from protein build-up and aggregation [45, 46] as is the case in neurodegenerative diseases [47]. Ongoing investigations will hopefully bring a more clear understanding of the molecular pathways that bortezomib targets in the near future.

Zangari and coworkers performed a retrospective analysis on three clinical trials in which bortezomib was given to patients [48]. Patients who had a partial response to bortezomib experienced a transient increase in alkaline phosphatase levels in their serum compared to nonresponders. This increase in alkaline phosphatase was found in all three trials. Similarly, Heider et al. reported that patients with relapsed MM enrolled in clinical trials using bortezomib had increased levels of bone-specific alkaline phosphatase and osteoclacin, markers of increased osteoblast activity, and that this increase in osteocalcin and alkaline phosphatase was observed in patients regardless if their MM responded to bortezomib [49]. Terpos et al. have reported that bortezomib monotherapy decreased the elevated bone resorption markers in patients with MM and decreased DKK1, an osteoblast inhibitor, and RANKL, an osteoclast stimulator. More importantly, bortezomib also increased bone formation markers in patients with relapsed myeloma [50]. However, in combination with melphalan, bortezomib could not enhance osteoblast function in patients [39]. The clearest evidence for bortezomib's effects on bone formation has been reported by Giuliani et al. [51]. In this report, the authors showed that bortezomib increased alkaline phosphatase and osteocalcin levels and enhanced osteoblast differentiation in human osteoblast precursor cultures as well as increased the critical osteoblast transcription factor, RUNX2. More importantly, bone marrow samples from patients responding to bortezomib showed increased numbers of osteoblastic cells and osteocalcin staining compared to nonresponders, and there was new bone formation clearly seen in the responding group. However, nonresponders did not show any effect of bortezomib on osteoblast differentiation. Similarly, Oyajobi and coworkers have shown that bortezomib also increases bone formation in murine models of MM, and that this effect can be inhibited by DKK1.

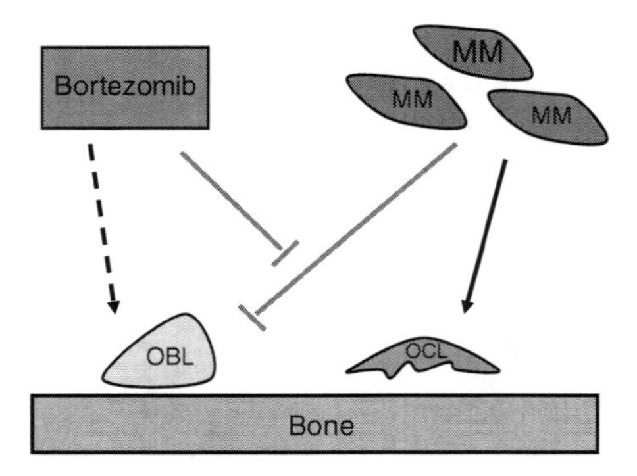

Fig. 1 Myeloma cells (MM) produce osteoclast (OCL) activating factors (see text for details) that increase osteoclast formation (*arrow*) as well as produce osteoblast (OBL) inhibiting factor (inhibitor). Bortezomib can enhance bone formation by increasing several osteoblast transcription factors (see text for details) (*dotted arrow*). In addition, bortezomib can inhibit osteoclast formation in addition to blocking growth of myeloma cells (inhibitor)

In addition, Qiang et al. [41] reported that bortezomib could induce osteoblast differentiation in MM patients independent of Wnt signaling. The mechanism that this group of investigators suggested was that bortezomib stabilized β-catenin and its nuclear translocation in a Wnt-independent manner.

Several groups have shown that increased osteoblast activity resulting from increased canonical Wnt signaling in the bone microenvironment can reduce tumor burden in preclinical models of MM [52, 53]. The mechanism responsible for increased osteoblast activity by bortezomib, however, has not been clearly defined. What has been shown by Garrett and coworkers [54] is that proteasome antagonists increased BMP2 production in the bone microenvironment. These data suggest that bortezomib, like other proteasome antagonists, may increase bone formation by increasing local BMP2 production. Thus, bortezomib appears to be a very appropriate agent for treating both MM as well as MM bone disease (see Fig. 1). However, there has only been one anecdotal report of bortezomib healing a lytic lesion in patients with MM. Japanese workers have reported that a skull lesion and a vertebral lesion of a patient on bortezomib were healed by computerized tomographic analysis. This has not been reported so far by other investigators.

4 Summary

Bortezomib is a first-in-class proteasome antagonist approved for the treatment of MM based on both very encouraging results from relapsed patients and newly diagnosed patients with MM. It appears to be active in patients who both have high risk MM with chromosome 13 deletions on karyotypic analysis and with poor prognostic markers. Bortezomib in combination with other bone anabolic agents or anti-bone resorptive agents may provide an excellent platform for treating both MM and MM bone disease. The encouraging findings of Giuliani et al. [51] and others who have suggested that bortezomib can have anabolic effects in patients with MM are extremely encouraging and suggest that bortezomib could contribute to the healing of lytic lesions in MM. However, bortezomib is frequently combined with agents that are toxic to bone such as dexamethasone and melphalan, so that the bone anabolic effect of bortezomib may be lost. For example, when bortezomib was combined with melphalan, which may have toxic effects on osteoblasts, patients no longer showed an anabolic response [55]. In addition to a bone formation stimulating activity, bortezomib also has been shown in preclinical models and in patients whose MM responded to bortezomib, to also inhibit bone resorption markers, and decrease osteoclast activity. Thus, bortezomib may have two effects on bone formation, both by decreasing bone destruction and by enhancing bone formation simultaneously. Future clinical trials combining bortezomib with agents such as denosumab, a RANKL antibody that is in clinical trial and appears to be effective in patients with osteoporosis and cancer treatment induced bone loss in prostate cancer [56, 57] or other targeted agents that are not cytotoxic to osteoblasts, may provide even greater benefits to patients with MM bone disease. The future is bright for patients with MM, whose median survival has increased significantly over the last 5 years with the introduction of newly targeted agents for treating MM in the MM microenvironment.

References

1. Roodman GD (2009) Pathogenesis of myeloma bone disease. Leukemia 23(3):435–441
2. Saad AA, Sharma M, Higa GM (2009) Treatment of multiple myeloma in the targeted therapy era. Ann Pharmacother 43(2):329–338
3. Mundy GR (1998) Myeloma bone disease. Eur J Cancer 34(2):246–251
4. Mundy GR, Raisz LG, Cooper RA, Schechter GP, Salmon SE (1974) Evidence for the secretion of an osteoclast stimulating factor in myeloma. N Engl J Med 291(20):1041–1046
5. Diamond T, Levy S, Day P, Barbagallo S, Manoharan A, Kwan YK (1997) Biochemical, histomorphometric and densitometric changes in patients with multiple myeloma: effects of glucocorticoid therapy and disease activity. Br J Haematol 97(3):641–648
6. Berenson JR, Lipton A (1998) Use of bisphosphonates in patients with metastatic bone disease. Oncology (Williston Park) 12(11):1573–1579, discussion 1579–1581
7. Boissy P, Andersen TL, Lund T, Kupisiewicz K, Plesner T, Delaisse JM (2008) Pulse treatment with the proteasome inhibitor bortezomib inhibits osteoclast resorptive activity in clinically relevant conditions. Leuk Res 32(11):1661–1668
8. Uy GL, Goyal SD, Fisher NM, Oza AY, Tomasson MH, Stockerl-Goldstein K, DiPersio JF, Vij R (2009) Bortezomib administered pre-auto-SCT and as maintenance therapy post transplant for multiple myeloma: a single institution phase II study. Bone Marrow Transplant 43(10):793–800
9. Jagannath S, Barlogie B, Berenson J, Siegel D, Irwin D, Richardson PG, Niesvizky R, Alexanian R, Limentani SA, Alsina M et al (2004) A phase 2 study of two doses of bortezomib in relapsed or refractory myeloma. Br J Haematol 127(2):165–172
10. Jagannath S, Barlogie B, Berenson JR, Siegel DS, Irwin D, Richardson PG, Niesvizky R, Alexanian R, Limentani SA, Alsina M et al (2008) Updated survival analyses after prolonged

follow-up of the phase 2, multicenter CREST study of bortezomib in relapsed or refractory multiple myeloma. Br J Haematol 143(4):537–540
11. Esteve FR, Roodman GD (2007) Pathophysiology of myeloma bone disease. Best Pract Res Clin Haematol 20(4):613–624
12. Calvani N, Silvestris F, Cafforio P, Dammacco F (2004) Osteoclast-like cell formation by circulating myeloma B lymphocytes: role of RANK-L. Leuk Lymphoma 45(2):377–380
13. Hjorth-Hansen H, Seifert MF, Borset M, Aarset H, Ostlie A, Sundan A, Waage A (1999) Marked osteoblastopenia and reduced bone formation in a model of multiple myeloma bone disease in severe combined immunodeficiency mice. J Bone Miner Res 14(2):256–263
14. Alsayed Y, Ngo H, Runnels J, Leleu X, Singha UK, Pitsillides CM, Spencer JA, Kimlinger T, Ghobrial JM, Jia X et al (2007) Mechanisms of regulation of CXCR4/SDF-1 (CXCL12)-dependent migration and homing in multiple myeloma. Blood 109(7):2708–2717
15. Hideshima T, Mitsiades C, Tonon G, Richardson PG, Anderson KC (2007) Understanding multiple myeloma pathogenesis in the bone marrow to identify new therapeutic targets. Nat Rev Cancer 7(8):585–598
16. Lentzsch S, Ehrlich LA, Roodman GD (2007) Pathophysiology of multiple myeloma bone disease. Hematol Oncol Clin North Am 21(6):1035–1049, viii
17. Giuliani N, Morandi F, Tagliaferri S, Colla S, Bonomini S, Sammarelli G, Rizzoli V (2006) Interleukin-3 (IL-3) is overexpressed by T lymphocytes in multiple myeloma patients. Blood 107(2):841–842
18. Hsu H, Lacey DL, Dunstan CR, Solovyev I, Colombero A, Timms E, Tan HL, Elliott G, Kelley MJ, Sarosi I et al (1999) Tumor necrosis factor receptor family member RANK mediates osteoclast differentiation and activation induced by osteoprotegerin ligand. Proc Natl Acad Sci USA 96(7):3540–3545
19. Boyle WJ, Simonet WS, Lacey DL (2003) Osteoclast differentiation and activation. Nature 423(6937):337–342
20. Matsuzaki K, Udagawa N, Takahashi N, Yamaguchi K, Yasuda H, Shima N, Morinaga T, Toyama Y, Yabe Y, Higashio K et al (1998) Osteoclast differentiation factor (ODF) induces osteoclast-like cell formation in human peripheral blood mononuclear cell cultures. Biochem Biophys Res Commun 246(1):199–204
21. Roodman GD (2007) Treatment strategies for bone disease. Bone Marrow Transplant 40 (12):1139–1146
22. Tsukii K, Shima N, Mochizuki S, Yamaguchi K, Kinosaki M, Yano K, Shibata O, Udagawa N, Yasuda H, Suda T et al (1998) Osteoclast differentiation factor mediates an essential signal for bone resorption induced by 1 alpha, 25-dihydroxyvitamin D3, prostaglandin E2, or parathyroid hormone in the microenvironment of bone. Biochem Biophys Res Commun 246 (2):337–341
23. Dougall WC, Glaccum M, Charrier K, Rohrbach K, Brasel K, De Smedt T, Daro E, Smith J, Tometsko ME, Maliszewski CR et al (1999) RANK is essential for osteoclast and lymph node development. Genes Dev 13(18):2412–2424
24. Sanz-Rodriguez F, Teixido J (2001) VLA-4-dependent myeloma cell adhesion. Leuk Lymphoma 41(3–4):239–245
25. Tai YT, Soydan E, Song W, Fulciniti M, Kim K, Hong F, Li XF, Burger P, Rumizen MJ, Nahar S et al (2009) CS1 promotes multiple myeloma cell adhesion, clonogenic growth, and tumorigenicity via c-maf-mediated interactions with bone marrow stromal cells. Blood 113 (18):4309–4318
26. Shi Y, Frost PJ, Hoang BQ, Benavides A, Sharma S, Gera JF, Lichtenstein AK (2008) IL-6-induced stimulation of c-myc translation in multiple myeloma cells is mediated by myc internal ribosome entry site function and the RNA-binding protein, hnRNP A1. Cancer Res 68(24):10215–10222
27. Abe M, Hiura K, Ozaki S, Kido S, Matsumoto T (2009) Vicious cycle between myeloma cell binding to bone marrow stromal cells via VLA-4-VCAM-1 adhesion and macrophage inflammatory protein-1alpha and MIP-1beta production. J Bone Miner Metab 27(1):16–23

28. Shain KH, Yarde DN, Meads MB, Huang M, Jove R, Hazlehurst LA, Dalton WS (2009) Beta1 integrin adhesion enhances IL-6-mediated STAT3 signaling in myeloma cells: implications for microenvironment influence on tumor survival and proliferation. Cancer Res 69 (3):1009–1015
29. Perez LE, Parquet N, Shain K, Nimmanapalli R, Alsina M, Anasetti C, Dalton W (2008) Bone marrow stroma confers resistance to Apo2 ligand/TRAIL in multiple myeloma in part by regulating c-FLIP. J Immunol 180(3):1545–1555
30. Kumatori A, Tanaka K, Tamura T, Fujiwara T, Ichihara A, Tokunaga F, Onikura A, Iwanaga S (1990) cDNA cloning and sequencing of component C9 of proteasomes from rat hepatoma cells. FEBS Lett 264(2):279–282
31. Grisham MB, Palombella VJ, Elliott PJ, Conner EM, Brand S, Wong HL, Pien C, Mazzola LM, Destree A, Parent L et al (1999) Inhibition of NF-kappa B activation in vitro and in vivo: role of 26S proteasome. Methods Enzymol 300:345–363
32. Adams J (2004) The proteasome: a suitable antineoplastic target. Nat Rev Cancer 4(5): 349–360
33. Landowski TH, Megli CJ, Nullmeyer KD, Lynch RM, Dorr RT (2005) Mitochondrial-mediated disregulation of Ca2+ is a critical determinant of Velcade (PS-341/bortezomib) cytotoxicity in myeloma cell lines. Cancer Res 65(9):3828–3836
34. Karin M, Lin A (2002) NF-kappaB at the crossroads of life and death. Nat Immunol 3 (3):221–227
35. Hideshima T, Nakamura N, Chauhan D, Anderson KC (2001) Biologic sequelae of interleukin-6 induced PI3-K/Akt signaling in multiple myeloma. Oncogene 20(42):5991–6000
36. Hideshima T, Chauhan D, Hayashi T, Akiyama M, Mitsiades N, Mitsiades C, Podar K, Munshi NC, Richardson PG, Anderson KC (2003) Proteasome inhibitor PS-341 abrogates IL-6 triggered signaling cascades via caspase-dependent downregulation of gp130 in multiple myeloma. Oncogene 22(52):8386–8393
37. Rajkumar SV, Kyle RA (2005) Multiple myeloma: diagnosis and treatment. Mayo Clin Proc 80(10):1371–1382
38. von Metzler I, Krebbel H, Hecht M, Manz RA, Fleissner C, Mieth M, Kaiser M, Jakob C, Sterz J, Kleeberg L et al (2007) Bortezomib inhibits human osteoclastogenesis. Leukemia 21 (9):2025–2034
39. Terpos E, Sezer O, Croucher P, Dimopoulos MA (2007) Myeloma bone disease and proteasome inhibition therapies. Blood 110(4):1098–1104
40. Pennisi A, Li X, Ling W, Khan S, Zangari M, Yaccoby S (2009) The proteasome inhibitor, bortezomib suppresses primary myeloma and stimulates bone formation in myelomatous and nonmyelomatous bones in vivo. Am J Hematol 84(1):6–14
41. Qiang YW, Hu B, Chen Y, Zhong Y, Shi B, Barlogie B, Shaughnessy JD Jr (2009) Bortezomib induces osteoblast differentiation via Wnt-independent activation of beta-catenin/TCF signaling. Blood 113(18):4319–4330
42. Silvestris F, Ciavarella S, De Matteo M, Tucci M, Dammacco F (2009) Bone-resorbing cells in multiple myeloma: osteoclasts, myeloma cell polykaryons, or both? Oncologist 14(3): 264–275
43. McConkey DJ (2009) Bortezomib paradigm shift in myeloma. Blood 114(5):931–932
44. Hideshima T, Chauhan D, Richardson P, Mitsiades C, Mitsiades N, Hayashi T, Munshi N, Dang L, Castro A, Palombella V et al (2002) NF-kappa B as a therapeutic target in multiple myeloma. J Biol Chem 277(19):16639–16647
45. Hideshima T, Chauhan D, Ishitsuka K, Yasui H, Raje N, Kumar S, Podar K, Mitsiades C, Hideshima H, Bonham L et al (2005) Molecular characterization of PS-341 (bortezomib) resistance: implications for overcoming resistance using lysophosphatidic acid acyltransferase (LPAAT)-beta inhibitors. Oncogene 24(19):3121–3129
46. Nawrocki ST, Carew JS, Maclean KH, Courage JF, Huang P, Houghton JA, Cleveland JL, Giles FJ, McConkey DJ (2008) Myc regulates aggresome formation, the induction of Noxa,

and apoptosis in response to the combination of bortezomib and SAHA. Blood 112(7): 2917–2926
47. Bennett EJ, Shaler TA, Woodman B, Ryu KY, Zaitseva TS, Becker CH, Bates GP, Schulman H, Kopito RR (2007) Global changes to the ubiquitin system in Huntington's disease. Nature 448 (7154):704–708
48. Zangari M, Esseltine D, Lee CK, Barlogie B, Elice F, Burns MJ, Kang SH, Yaccoby S, Najarian K, Richardson P et al (2005) Response to bortezomib is associated to osteoblastic activation in patients with multiple myeloma. Br J Haematol 131(1):71–73
49. Heider U, Kaiser M, Muller C, Jakob C, Zavrski I, Schulz CO, Fleissner C, Hecht M, Sezer O (2006) Bortezomib increases osteoblast activity in myeloma patients irrespective of response to treatment. Eur J Haematol 77(3):233–238
50. Terpos E, Heath DJ, Rahemtulla A, Zervas K, Chantry A, Anagnostopoulos A, Pouli A, Katodritou E, Verrou E, Vervessou EC et al (2006) Bortezomib reduces serum dickkopf-1 and receptor activator of nuclear factor-kappaB ligand concentrations and normalises indices of bone remodelling in patients with relapsed multiple myeloma. Br J Haematol 135(5):688–692
51. Giuliani N, Morandi F, Tagliaferri S, Lazzaretti M, Bonomini S, Crugnola M, Mancini C, Martella E, Ferrari L, Tabilio A et al (2007) The proteasome inhibitor bortezomib affects osteoblast differentiation in vitro and in vivo in multiple myeloma patients. Blood 110 (1):334–338
52. Edwards CM (2008) Wnt signaling: bone's defense against myeloma. Blood 112(2):216–217
53. Qiang YW, Chen Y, Stephens O, Brown N, Chen B, Epstein J, Barlogie B, Shaughnessy JD Jr (2008) Myeloma-derived Dickkopf-1 disrupts Wnt-regulated osteoprotegerin and RANKL production by osteoblasts: a potential mechanism underlying osteolytic bone lesions in multiple myeloma. Blood 112(1):196–207
54. Garrett IR, Chen D, Gutierrez G, Zhao M, Escobedo A, Rossini G, Harris SE, Gallwitz W, Kim KB, Hu S et al (2003) Selective inhibitors of the osteoblast proteasome stimulate bone formation in vivo and in vitro. J Clin Invest 111(11):1771–1782
55. Terpos E (2008) Bortezomib directly inhibits osteoclast function in multiple myeloma: implications into the management of myeloma bone disease. Leuk Res 32(11):1646–1647
56. Smith MR, Egerdie B, Hernandez Toriz N, Feldman R, Tammela TL, Saad F, Heracek J, Szwedowski M, Ke C, Kupic A et al (2009) Denosumab in men receiving androgen-deprivation therapy for prostate cancer. N Engl J Med 361(8):745–755
57. Cummings SR, San Martin J, McClung MR, Siris ES, Eastell R, Reid IR, Delmas P, Zoog HB, Austin M, Wang A et al (2009) Denosumab for prevention of fractures in postmenopausal women with osteoporosis. N Engl J Med 361(8):756–765

Bortezomib in the Upfront Treatment of Multiple Myeloma

Jesús F. San Miguel and María -Victoria Mateos

Abstract Although multiple myeloma (MM) remains a disease that is incurable with conventional treatments, important changes have recently been introduced in the management of the disease as a result of advances in our knowledge of its pathogenesis and the availability of novel agents. Bortezomib is a first-in-class proteasome inhibitor that targets not only the MM cell but also its interaction with the bone marrow microenvironment. It represents an excellent example of a novel class of agents that have quickly moved from the bench to the bedside. Four randomised trials have evaluated the role of bortezomib-based combinations as an induction therapy in transplant candidate myeloma patients, revealing a high efficacy (>80% response rate, with 20–30% complete response (CR)) that increased after autologous stem cell transplant, confirming the results of numerous pilot studies of various bortezomib-based combinations. In patients who are not candidates for transplant, bortezomib in combination with melphalan and prednisone has also proved to be superior to conventional therapy, with high overall and CR rates and a significantly longer time to progression and overall survival. Moreover, new strategies are being explored in patients who are not candidates for transplant based on the optimisation of the VISTA schedule and using weekly doses of bortezomib to improve tolerability. Overall, these results have established bortezomib-based combinations as key treatment options for newly diagnosed myeloma patients.

J.F. San Miguel (✉)
University Hospital of Salamanca, Paseo San Vicente 58-182, 37007 Salamanca, Spain
CIC, IBMCC (USAL-CSIC), Campus Unamuno s/n, 37007 Salamanca, Spain
e-mail: sanmigiz@usal.es

M.-V. Mateos
University Hospital of Salamanca, Paseo San Vicente 58-182, 37007 Salamanca, Spain

I.M. Ghobrial et al. (eds.), *Bortezomib in the Treatment of Multiple Myeloma*,
Milestones in Drug Therapy,
DOI 10.1007/978-3-7643-8948-2_4,

1 Introduction

Multiple myeloma (MM) is the second most common haematological malignancy after non-Hodgkin's lymphoma. Estimates suggest that 20,580 new cases of MM will be diagnosed in the USA in 2009, with an estimated 10,580 deaths (1). Clinical features and symptoms arise from clonal proliferation of myeloma cells within the bone marrow. These include bone destruction, which leads to fractures, disruption of normal bone marrow and immune function, which cause anaemia and increased susceptibility to infections, hypocalcaemia, which is a consequence of increased bone resorption, hyperviscosity, which is related to high serum levels of monoclonal proteins secreted by myeloma cells,and impaired renal function.

Historically, treatment options for young, newly diagnosed patients included cytoreductive regimens (such as VAD: vincristine, adryamycin and dexamethasone) followed by high-dose therapy (usually melphalan, 200 mg/m^2) plus stem cell transplantation (HDT-SCT), while MP (melphalan plus prednisone) remained as the standard of care for patients who were not candidates for HDT-SCT. However, it is important to note that VAD as induction regimen resulted in a complete remission (CR) rate of $<$10%, while MP in elderly patients yielded CR in only approximately 2% of cases (2, 3). Therefore, the treatment options for MM patients were limited, and in most cases, the outcome was very disappointing. There was an urgent need for more active therapies that could improve the outcome of MM patients. Over the last 10 years, three new drugs have emerged to treat them: thalidomide (Thalidomide®, Celgene Corp), lenalidomide (Revlimid®, Celgene Corp) and bortezomib (VELCADE®, Millennium Pharmaceuticals: The Takeda Oncology Company and Johnson & Johnson Pharmaceutical Research & Development, L.L.C.). These novel drugs have shown themselves to be effective in relapsed and refractory patients and are currently being evaluated as front-line therapy.

Bortezomib, a first-in-class proteasome inhibitor, is currently approved in the USA and the European Union (EU) for the treatment of MM patients who have received at least one previous therapy. It has recently been approved for previously untreated MM patients in the USA and in combination with melphalan and prednisone for the treatment of patients with previously untreated MM who are not eligible for HDT/SCT in the EU. This review focuses on the results obtained with bortezomib and bortezomib-based combinations in previously untreated patients, both as induction therapy in patients eligible for HDT-SCT and in non-transplant candidate patients.

2 Bortezomib as Monotherapy

Bortezomib as monotherapy, at the standard dose (1.3 mg/m^2) and with a conventional schedule (days 1, 4, 8 and 11 followed by a 10-day rest period), has been explored in a multicentre, phase II trial of 64 patients. The overall response rate

(ORR) was 41%, including 9% complete remission (CR)/near-CR (4); the median time to progression (TTP) was 17.3 months and estimated 1-year survival was 92%. While 64% of patients had treatment-emergent peripheral neuropathy, grade 3 or higher was observed in only 3% of them; the toxicity was manageable and resolved in 85% of cases within a median of 98 days. Notably, neurological examination, including neurophysiologic tests and evaluations of epidermal nerve fibre densities showed pre-treatment sensory neuropathy in 20% of patients and new or worsening neuropathy in 63%. Another phase II trial with single-agent bortezomib was conducted by Dispenzieri et al. in which the same scheme of bortezomib as monotherapy was examined in 42 untreated patients with high-risk characteristics. The ORR was 45%, with no differences among subgroups of patients with different characteristics of poor prognosis (elevated plasma cell labelling index, elevated beta-2 microglobulin, translocation t(4;14) or chromosome 13 deletion) (5). Nevertheless, 25% of these high-risk patients showed early progression, suggesting that although bortezomib is an effective drug in high-risk MM patients, either consolidation therapy or combination with other agents would be more effective.

3 Bortezomib-based Combinations in Transplant Candidate MM Patients

3.1 Randomised Phase III Trials

The role of bortezomib-based combinations as induction therapy in transplant candidates MM patients has been evaluated in four randomised phase III trials (Table 1). The ultimate aim of all of them was to evaluate whether these bortezomib-based combinations could replace the use of conventional debulky schemes by yielding higher response rates than conventional regimens, before and after autologous stem cell transplant (ASCT), a prolongation of progression-free survival (PFS) and, ultimately, of overall survival (OS). Nevertheless, it should be noted that the impact on OS is difficult to demonstrate, since the heterogeneity of treatments at relapse precludes a statistically sound analysis.

The Intergroupe Français du Myélome (IFM) has initiated a randomised trial in which standard-dose bortezomib plus dexamethasone is compared with the conventional VAD regimen in 482 MM HDT/SCT candidate patients. The primary objective is to determine the CR rate (CR + near-CR) after four cycles. Data presented at the ASH/ASCO symposium of the 2008 ASH meeting showed that bortezomib–dexamethasone was clearly superior to VAD in terms of partial response (PR) (82 vs. 65%) and CR + near-CR (15 vs. 7%) and, most importantly, that the superiority of this combination was maintained after HDT/SCT, reaching a CR + near-CR of 40 vs. 22%, and very good partial response (VGPR) or better of 61 vs. 44%. These efficacies were translated into a clinical benefit for patients, as fewer of them required a second SCT (6). The superiority of bortezomib–dexamethasone as induction was observed in all prognostic subgroups, regardless of ISS stage,

Table 1 Randomised trials with bortezomib-based combinations in newly diagnosed MM HDT/SCT-candidate patients

Author	Patients	Regimen	Post-induction response rate (≥PR)	Post-transplant response rate (≥PR)
Harousseau	485	Bortezomib plus dexamethasone vs. VAD	82 vs. 65% (15 vs. 7% CR/near-CR)	91 vs. 91% (40 vs. 22% CR/near-CR)
Sonneveld	300	Bortezomib plus adriamycin and dexamethasone (PAD) vs. VAD	79 vs. 57% (7 vs. 2% CR/near-CR)	91 vs. 79% (26 vs. 14% CR/near-CR)
Cavo	256	Bortezomib plus thalidomide plus dexamethasone vs. thalidomide plus dexamethasone	94 vs. 79% (32 vs. 12% CR/near-CR)	76 vs. 88% ≥VGPR (55 vs. 32% CR/near-CR)
Rosiñol	183	VBCMP/VBAD plus bortezomib vs. bortezomib plus thalidomide plus dexamethasone vs. thalidomide plus dexamethasone	72 vs. 80 vs. 66% (28 vs. 41 vs. 12% CR/near-CR)	97 vs. 97% vs. 97% (54 vs. 64 vs. 53% CR/near-CR)

increased β2-microglobulin, del[13], t(4;14) and/or del[17p]. This trial has also explored the value of two consolidation DCEP cycles before HDT/SCT, showing that this strategy apparently confers no benefit since the CR rate did not increase after DCEP. In terms of survival, a significant prolongation of PFS has already been observed for bortezomib–dexamethasone compared with VAD (69 vs. 60% at 2 years, respectively; $p = 0.01$), while no significant differences in OS have so far been detected.

The synergistic effect observed in preclinical studies when bortezomib was combined with doxorubicin as well as its efficacy in relapsed/refractory patients was the rationale for the ongoing trial conducted by the HOVON group that compared the combination of bortezomib–dexamethasone plus adriamycin (PAD) vs. VAD. After three induction cycles, PAD induced a significantly higher VGPR rate (45 vs. 17%) and after HDT/SCT, the CR rate was also significantly higher in the PAD arm (26 vs. 14%; $p = 0.003$). Patients with del[13/13q] had a superior response with PAD than with VAD; nevertheless, the presence of t(4;14) adversely affected the outcome of both schedules (7).

The combination of bortezomib with immunomodulatory drugs (IMiDs) is being explored by the GIMEMA group in a phase III trial in which 460 patients have been randomised to receive either bortezomib–thalidomide–dexamethasone (VTD) or thalidomide–dexamethasone (TD) followed by tandem transplant. Results of this study have revealed a significant superiority of VTD as induction regimen; after three VTD induction cycles, 21% of patients achieved CR compared with 6% with TD, and the first HDT-SCT improved these efficacy results and maintained the superiority of VTD as measured by CR (43 vs. 23%) and VGPR or better (76 vs. 58%). Responses to VTD were rapid, with an acceptable and manageable safety profile. The superiority of VTD over TD was observed in low-risk and high-risk subgroups, including those with cytogenetic abnormalities such as del[13], t(4;14) and del[17p] (8). After a median follow-up of 15 months, the 2-year PFS was already significantly superior for VTD (90 vs. 80%; $p = 0.009$) but so far without improved OS.

The PETHEMA/GEM group (Spanish Myeloma Group) is conducting a randomised trial also comparing VTD with TD but with six induction cycles and the introduction of a third arm with four cycles of VBCMP/VBAD on an alternating basis followed by two cycles of bortezomib at a standard dose and with a conventional scheme. Data from an interim analysis of 183 patients indicated that VTD and VBCMP/VBAD were better than TD. The superiority of VTD and chemotherapy plus bortezomib over TD was maintained after HDT/SCT, with CR rates of 49 and 43%, respectively, compared with 34% for TD (9). This interim analysis showed a trend towards better responses in the patients with high-risk of cytogenetic abnormalities who received VTD compared with those in the other two arms.

Overall, the most frequently reported side effects with the front-line use of bortezomib have been manageable and comparable to those reported in relapsed patients. Nevertheless, the incidence was lower because only three or four induction cycles were administered. The most commonly reported adverse events were gastrointestinal events (27% for all grades), haematological toxicity (G3-4 neutropenia, 5%, G3-4 thrombocytopenia, 3%), fatigue (28% for all grades) and, particularly, peripheral neuropathy. The frequency of G3-4 peripheral neuropathy was 6–7% and was reversible in most cases. Nevertheless, rigorous monitoring of peripheral neuropathy and the use of established dose-modification guidelines is highly recommended.

3.2 *Pilot Studies*

The results obtained in the randomised trials described above are consistent with those of previous pilot studies in which the role of bortezomib-based combinations was first evaluated in the setting of this patient population (Table 2).

3.2.1 Bortezomib Plus Dexamethasone

Five phase II studies have evaluated the role of bortezomib plus dexamethasone as induction therapy in young, newly diagnosed MM patients. Jagannath et al. conducted a trial in which patients who did not respond favourably to single-agent bortezomib were permitted to add dexamethasone (40 mg) on the day of each bortezomib dose and the day after. The best ORR was 90%, including 19% CR/near-CR cases. The addition of dexamethasone to bortezomib improved the response in 77% of patients, the majority being improvements featuring transitions from either stable disease or minor response to PRs. Median OS had not been achieved and the 4-year estimated OS was 67% at the time of publication (10).

Table 2 Pilot studies with bortezomib-based combinations in young, newly diagnosed MM patients

Author	Patients	Regimen	Post-induction response rate (≥PR)	Post-transplant response rate (≥PR)
Bortezomib single-agent/bortezomib plus dexamethasone				
Richardson	64	Bortezomib single-agent	41% (9% CR)	–
Dispenzieri	42	Bortezomib single-agent	45%	–
Jagannath	49	Bortezomib single-agent +/– dexamethasone	44% (single-agent) 88% (18% CR) (+ dex)	–
Harousseau	52	Bortezomib plus dexamethasone	67% (21% CR)	90% (33% CR)
Corso	54	Bortezomib plus dexamethasone	86% (50% CR)	94% (56% CR)
Na Nakorn	20	Bortezomib plus dexamethasone	79% (37% CR)	–
Rosiñol	40	Bortezomib/dexamethasone (alternating cycles)	65% (13% CR)	88% (33% CR)
Bortezomib plus anthracyclines				
Oarkowee	21	Bortezomib (1.3 mg/m^2) plus doxorubicin plus dexamethasone	95% (29% CR)	95% (57% CR)
Popat	20	Bortezomib (1.0 mg/m^2) plus doxorubicin plus dexamethasone	89% (16% CR)	89% (42%)
Orlowski	29	Bortezomib plus pegLD	79% (28% CR/near-CR)	–
Jakubowiak	40	Bortezomib plus pegLD plus dexamethasone	93% (40% CR)	87% (57% CR)
Belch	50	Bortezomib plus pegLD plus dexamethasone	78% (18% CR)	93% (27% CR)
Palumbo	102	Bortezomib plus pegLD plus dexamethasone	94% (13% CR)	88% ≥VGPR (41% CR)
Bortezomib plus alkylating agents				
Kropff	30	Bortezomib plus cyclophosphamide plus dexamethasone	77% (10% CR)	–
Reeder	33	Bortezomib plus cyclophosphamide plus dexamethasone	88% (39% CR/near-CR)	100% (70% CR/near-CR)
Berenson	35	Bortezomib plus melphalan plus ascorbic acid	74% (16% CR)	–
Bortezomib plus immunomodulatory drugs				
Bortezomib plus thalidomide				
Borrello	27	Bortezomib plus thalidomide	82% (22% CR)	–
Chen	30	Bortezomib plus thalidomide	89% (50% CR)	–
Wang	38	Bortezomib plus thalidomide plus dexamethasone	87% (16% CR)	95% (37% CR)
Kaufman	44	Bortezomib plus thalidomide plus dexamethasone	91% (23% CR)	100% (47% CR)
Giralt	27	Bortezomib plus thalidomide plus dexamethasone	95% (15% CR)	66% CR
Badros	12	VTD-PACE	83% (17% CR/near-CR)	92% (58% CR)
Barlogie	303	VTD-PACE	–	82% CR/near-CR at 4 years
Bortezomib plus lenalidomide				
Richardson	68	Bortezomib plus lenalidomide plus dexamethasone	98% (36% CR/near-CR)	–
Kumar	25	Bortezomib, lenalidomide, cyclophosphamide plus dexamethasone	100% (32% CR/near-CR)	–

Bortezomib at a standard dose combined with dexamethasone as induction therapy for HDT/SCT was explored by the IFM (11) in a trial in which 48 patients received four cycles of standard-dose bortezomib plus dexamethasone. ORR after induction therapy was 66%, including 21% CR and 10% VGPR. Response rates improved and the CR plus VGPR rate increased from 31 to 54% after HDT/SCT. Corso et al. reported an ORR of 86%, including 50% CR/near-CR and 70% ≥VGPR, in 54 MM patients after four cycles of bortezomib plus dexamethasone (12). Using the same scheme, Na Nakorn et al. reported an ORR of 79%, including 37% CR, in 20 patients (13). Finally, this combination has also been explored by the PETHEMA/GEM group, but instead of giving both drugs concomitantly, they used six alternating cycles of bortezomib and dexamethasone as induction therapy before HDT-SCT (14). Forty patients were included and the ORR was 60%, including 13% CR; this response rate was also upgraded following HDT-SCT (ORR of 88% with 33% CR). This combination was associated with very low toxicity and enabled excellent stem cell harvesting.

3.2.2 Bortezomib Plus Anthracyclines

The PAD regimen, consisting of the combination of bortezomib at standard dose and with a conventional schedule, plus dexamethasone and escalating doses of doxorubicin (0–9 mg/m^2), was investigated in a series of 21 newly diagnosed SCT candidate patients (15). After PAD induction, 20 of the 21 patients (95%) achieved at least PR, including CR in five patients (24%). Eighteen of 21 patients received HDT-SCT and the rate of CR/near-CR increased from 29% after PAD to 57% after SCT. These results have been updated to show a median PFS time of 29 months, a median time to next therapy (TNT) of 36 months and a 2-year OS rate of 95%. Peripheral neuropathy was the most frequent adverse event, being present in 48% of the patients. In an attempt to alleviate this side effect, the same group included a subsequent cohort of 19 patients in which the dose of bortezomib was reduced to 1.0 mg/m^2; a similar efficacy in the response rate was achieved (89% ORR, with 16% CR + near-CR), with a slightly shorter median PFS (24 months), TNT (29 months) and 2-year OS rate (73%), although grade 3–4 peripheral neuropathy was not present (16).

This same combination, but using pegylated liposomal doxorubicin (pegLD) (30 mg/m^2 on day 4) and a standard dose of bortezomib and dexamethasone (20 mg on day 1 and after each bortezomib dose) has been explored in two phase II trials (17, 18), which confirmed the high efficacy for this schedule as induction therapy. CR/near-CR rates of up to 43%, increasing to 65% after HDT/SCT were reported. It is also of note that the Italian group used this combination prior to reduced-intensity HDT/SCT in elderly MM patients and obtained an ORR of 94%, including 21% CR/near-CR, which increased to 59% after SCT, and a 2-year OS of 92% (19). A regimen combining bortezomib plus pegLD without steroids was also investigated by Orlowski et al. in 63 patients. They reported an ORR of 79 and 28% CR/near-CRs (20).

3.2.3 Bortezomib Plus Alkylating Agents

Cyclophosphamide has been the alkylating agent most frequently used in combination with bortezomib in young, newly diagnosed MM patients, and it has produced promising results in two trials. One phase I trial evaluated bortezomib plus dexamethasone in combination with escalating doses of intravenous cyclophosphamide in 30 patients. The maximum tolerated dose of cyclophosphamide was defined as 900 mg/m^2 and at this dose the ORR was 92% (21). The same schedule, but using a fixed dose of oral cyclophosphamide (300 mg/m^2 on days 1, 8, 15 and 22) was analysed in a phase II trial of 33 patients. The ORR was 88% after four induction cycles, including 39% of CR/near-CR. All patients received HDT/SCT and the CR/near-CR was upgraded to 70% (22). Melphalan, instead of cyclophosphamide, was also combined with oral ascorbic acid plus bortezomib, at a dose of 1.0 mg/m^2 in a steroid-free scheme in 35 patients, giving rise to an ORR of 74%, including 26% of CR/near-CR (23).

3.2.4 Bortezomib in Combination with Immunomodulatory Drugs

Bortezomib Plus Thalidomide-based Combinations

A regimen using bortezomib plus thalidomide, without steroids, was examined in two phase II trials of 27 and 30 patients. The ORRs were 82 and 89%, respectively, including up to 50% of CR/near-CR rates in the Chinese study (24, 25).

The efficacy of the combination of bortezomib–thalidomide plus dexamethasone (VTD) was analysed in three pilot studies. Wang et al. explored increasing doses of bortezomib (1.0–1.6 mg/m^2), thalidomide (100–200 mg) and dexamethasone in a series of 38 patients. They reported an ORR of 87%, with 16% CR This result was not influenced by the dose of bortezomib (26). Similar results were obtained in two other studies. In the first, conducted by Giralt et al. (27), VTD was superior to thalidomide plus dexamethasone (TD) alone (CR rate pre-HDT-SCT, 15 vs. 6% and post-HDT-SCT, 38 vs. 23%). Similar efficacies were reported in the second retrospective study conducted by Kaufman et al. (28). All these trials showed that time to response to VTD was less than 1.5 months, so that only two cycles of the regimen were necessary to prepare the patients for subsequent HDT-SCT.

A more complicated scheme was employed by Badros et al. to determine the maximum tolerated dose of bortezomib (dose escalation from 0.7 to 1.3 mg/m^2) in combination with DT-PACE (cisplatin, doxorubicin, cyclophosphamide, etoposide, dexamethasone and thalidomide) (29). Twelve patients completed the study and all received HDT/SCT. After two cycles, the ORR was 83%, with 17% CR/near-CR. This scheme was subsequently evaluated in the Total Therapy 3 programme (TT3) as induction and consolidation therapy post-tandem transplant, plus VTD as maintenance therapy (30). The 4-year CR and near-CR rates were 87 and 78%, respectively, with 4-year OS and EFS rates of 78 and 71%, respectively. When these results were compared with those of the Total Therapy 2 programme, it was found that the addition of bortezomib improved CR duration and EFS (with a trend towards better OS) (31). Nevertheless, the superiority of the TT3 programme was mainly observed in the low-risk subgroup of patients, while in the high-risk MM group, the outcomes were also poor with TT3.

Bortezomib Plus Lenalidomide-based Combinations

The combination of bortezomib with lenalidomide and dexamethasone was investigated in a phase I/II trial of 68 enrolled patients. All patients responded, including 74% ≥VGPR and 44% CR/near-CR. Moreover, responses were independent of cytogenetics and toxicities were manageable, with an incidence of only 3% of grade 3 peripheral neuropathy and deep venous thrombosis (DVT) (32). Kumar et al. explored the same combination but with the addition of cyclophosphamide in 25 patients. All of them responded, with 20% stringent CR and an additional 16% CR (33). Taken together, these data suggest that the upfront combination of a proteasome inhibitor plus one IMID is highly effective, although longer follow-up is needed to determine the impact on survival and, in particular, to exclude the possibility of inducing more resistant relapses and avoiding burning out drugs that could be very valuable at relapse.

3.3 Impact of Bortezomib-based Combinations on Stem Cell Collection

As mentioned previously, bortezomib does not have an adverse impact on stem cell collection or post-SCT engraftment. This has been observed in all previously described trials in which bortezomib-based combinations were used as an induction therapy prior to HDT-SCT. In addition, bortezomib was also incorporated as part of a mobilisation approach with cyclophosphamide and granulocyte colony-stimulating factor (G-CSF), resulting in a median stem cell collection of 21×10^6 CD34+ cells/kg (34).

3.4 Bortezomib as Part of the Conditioning Regimen for SCT

Based on the synergistic effect of bortezomib and melphalan demonstrated in in vitro and in vivo studies, as well as the different toxicity profile of these two drugs, the possibility of adding bortezomib to the conditioning regimen has also been investigated. The IFM Group conducted a pilot study of Mel200 plus bortezomib (1 mg/m^2 on days −6, −3, +1 and +4) (35). Three months after SCT, 70% of patients had achieved ≥VGPR, including 34% with CR. Kaufman et al. (36)

conducted another pilot study of Mel200 plus bortezomib (increased from 1.0 to 1.6 mg/m^2, administered at single dose 24 h before or after Mel200) in which the PR rate was 94%, with 53% VGPR.

3.5 *Role of Bortezomib as Consolidation/Maintenance Therapy*

The use of bortezomib-based combinations as consolidation therapy has been tested with the aim of improving the quality of responses by converting patients in PR into CR. The GIMEMA group ran a trial in which VTD was administered as post-HDT/SCT consolidation therapy. It achieved an improvement in the CR/near-CR rate from 36 to 81% and, most importantly, after four 5-week cycles of VTD, six out of 29 patients had achieved true molecular remission. Strikingly, no clinical relapses were observed in these patients after a median follow-up of 26 months (37). Several current trials are also evaluating the role of bortezomib as maintenance therapy. Preliminary results from the HOVON group, in which bortezomib was administered every 15 days after HDT/SCT, showed an improvement in CR rate from 23 to 37% (7). The PETHEMA/GEM group is exploring the administration of a conventional cycle of bortezomib every 3 months in combination with continuous low dose of thalidomide up to 3 years as maintenance therapy after HDT/SCT (9).

4 Bortezomib in Non-HDT/SCT Candidate Patients

At the time of diagnosis, half of the MM patients are $\geq$65 years old and therefore not transplant candidates. MP has been used as the standard therapy since the 1960s, but with disappointing results (ORR of 45–60%, with rare CR, median PFS time of 18 months and median OS time of 29 months) (3). Thus, new treatment strategies were urgently needed for these patient populations. Table 3 summarises the trials conducted in MM patients who were not candidates for transplant that employed bortezomib-based regimens. Mateos et al. (38) examined the efficacy of adding bortezomib to the conventional MP scheme (melphalan at a dose of 9 mg/m^2 and prednisone at a dose of 60 mg/m^2 for 4 days) (VMP) in elderly, untreated MM patients. Following the phase I study in which 1.3 mg/m^2 was defined as the appropriate dose for bortezomib in this combination, 60 patients were recruited, half of whom were $\geq$75 years. The ORR was 89%, with 32% CR plus 11% near-CR; the median TTP was 27 months and the median OS was 58 months (unpublished data). These results led to the international, randomised, phase III VISTA (Velcade as Initial Standard Treatment: Assessment with melphalan and prednisone) trial in which 685 patients were randomised to receive either VMP or MP alone (39). Results of this trial confirmed the superiority of VMP over MP for all response and time-to-event measures. According to the stringent EBMT criteria, VMP resulted in a CR rate of 30%, which was significantly superior to that obtained in

Table 3 Bortezomib-based combinations in patients who are not candidates for HDT/SCT MM

Author	Patients	Regimen	Response rate (≥PR)
Mateos	60	Bortezomib plus melphalan and prednisone	89% (32% CR)
San Miguel	682	Bortezomib plus melphalan and prednisone vs. melphalan and prednisone	71 vs. 35% (30 vs. 4% CR)
Mateos[a]	260	Bortezomib plus melphalan and prednisone vs. bortezomib plus thalidomide and prednisone	81 vs. 81% (22 vs. 27% CR)
Palumbo[a]	354	Bortezomib plus melphalan and prednisone vs. bortezomib, melphalan, thalidomide and prednisone	45 vs. 55% ≥VGPR (16 vs. 31% CR)

[a]Schedules based on the use of weekly administration of bortezomib

the MP arm (4%). Responses to VMP were rapid and durable; the median time to response was 1.4 months and median duration of response was 19.9 months. The superiority of VMP was observed also in terms of TTP (median: 24 vs. 16 months) and OS; with an updated median follow-up of 26 months, the OS analysis showed a 36% reduction in the risk of death for VMP with a 3-year OS of 72 and 59% for VMP and MP, respectively (40). Efficacy of VMP was also evaluated in three poor prognosis subgroups. In patients older than 75 years, the median TTP was identical to that of younger patients, the CR rate was slightly lower (26 vs. 32%) and there was a trend towards shorter OS. With respect to renal impairment, response rate and TTP did not appear to be significantly different in patients with glomerular filtration rates of ≤50 or >50 mL/min and OS appeared somewhat longer in patients with normal renal function (41). Finally, with respect to cytogenetic abnormalities, there were no significant differences in CR rate, median TTP and OS between patients at high-risk (t(4;14), t(14, 16) and/or del[17]) and standard-risk of cytogenetic abnormalities. Interestingly, novel parameters reflecting a clinical benefit, such as TNT and treatment-free interval (TFI) also argue in favour of VMP. Median TNT and TFI were significantly longer for the VMP arm than with MP (28.1 and 16.1 months vs. 20.8 and 9.8 months, respectively). Fewer patients in the VMP arm (38%) than in the MP arm (57%) required subsequent therapy. Of those who received subsequent therapy in the VMP and MP arms, respectively, 43 (24%) and 116 (50%) patients received bortezomib, 81 (46%) and 110 (47%) received thalidomide, and 57 (32%) and 30 (13%) received lenalidomide. Response rates to second-line bortezomib-, thalidomide- and lenalidomide-based therapies were 41, 37 and 73%, respectively, following VMP, and 59, 47 and 67%, respectively, following MP (40). These findings suggest that patients who relapsed after VMP were not more intrinsically resistant than after using a traditional MP regimen. In terms of toxicity, the most divergent grade 3–4 toxicities between VMP and MP were gastrointestinal events (20 vs. 6% in VMP and MP, respectively) and peripheral neuropathy (13% in VMP and 0% in MP). In addition, 17 and 14% of VMP patients experienced grade 2 and 1 peripheral neuropathy, respectively, contributing to an overall incidence of 44%. Nevertheless, this adverse event was resolved in 56% of patients in a median of 5.7 months and improved in 18% of them after a median of 2 months.

In an attempt to optimise the treatment of HDT/SCT non-candidate patients, the PETHEMA/GEM and GIMEMA groups carried out randomised trials focused on the use of less-intensive VMP schemes in order to try to maintain efficacy and decrease toxicity. The Spanish trial (GEM05MAS65) was designed to compare six cycles of induction with VMP and VTP in 260 patients. It differed from the VISTA trial, in that the VMP regimen was based on only a single-intensive 6-week cycle (bortezomib administered on days 1, 4, 8, 11, 22, 25, 29 and 32) although in the subsequent five cycles, bortezomib was given only once a week. The VTP arm featured the same conditions as VMP except that melphalan was substituted by thalidomide. Preliminary results showed no significant differences in efficacy (81% ORR in both arms, with CR rates of 22 and 27% for VMP and VTP, respectively) (42). Haematological toxicity was slightly higher for VMP than for VTP, leading to a higher proportion of grade 3 infections in this arm (7 vs. <1%); by contrast, grade 3 cardiac events were more frequent in the VTP arm (0 vs. 8.5%). With respect to peripheral neuropathy, the VMP and VTP induction regimens had an incidence of 5 and 9%, respectively. These figures argue in favour of these modified schedules compared with the VISTA trial. Finally, in this trial, after induction therapy, patients were randomised to receive maintenance therapy for 3 years with bortezomib at the standard dose and with a conventional schedule every 3 months plus a low dose of either thalidomide or prednisone. In the Italian trial, VMP was compared with VMPT. Patients initially received a scheme of treatment similar to that previously reported in the VISTA trial (bortezomib administered twice a week), with thalidomide being added in the VMPT arm. In March 2007, the protocol was amended and both VMP and VMPT schedules were changed to nine 5-week cycles in which bortezomib was given only once a week. Preliminary results reported when the response of 354 patients could be evaluated showing a significantly higher ≥VGPR rate in the VMPT than in the VMP arm (55 vs. 45%), including a CR rate of 31% for VMPT and 16% for VMP (43). Subgroup analysis revealed no statistically significant differences in responses, ISS stage or chromosomal abnormalities (t(4;14), t(14;16) or del[17]) between the two groups. There were no differences in haematological toxicity between the VMPT and VMP arms (neutropenia: 36 and 31%; thrombocytopenia: 20 and 19%, respectively). Finally, grade 3–4 peripheral neuropathy was observed in 24 and 14% of patients receiving VMPT and VMP, respectively, at the twice weekly schedule, but it dropt to 6 and 3% with the weekly scheme.

5 Conclusions

Results to date indicate that bortezomib is a valuable drug for newly diagnosed patients, particularly in combination with dexamethasone, alkylating agents, anthracyclines and immunomodulatory drugs. In younger patients who are candidates for transplant, four randomised phase III clinical trials have demonstrated that bortezomib–dexamethasone, bortezomib–adriamycin–dexamethasone and bortezomib–dexamethasone–thalidomide are superior to the standard regimens of VAD and thalidomide plus dexamethasone. The benefit is maintained after HDT/SCT and it already translates into a prolonged PFS. Bortezomib does not affect stem cell collection or subsequent engraftment. In patients who are not candidates for high-dose therapy, VMP was able to prolong OS relative to MP alone in the VISTA trial and was superior for all the pre-specified efficacy end points, including TTP, TNT and TFI, which confirms that VMP is a new gold standard of care for this patient population. Preliminary results from ongoing modified VMP schemes using bortezomib at weekly intervals suggest that the toxicity can be reduced while maintaining efficacy.

References

1. Jemal A, Siegel R, Ward E, Hao Y, Xu J, Thun MJ (2009) Cancer statistics, 2009. CA Cancer J Clin 59:225–249
2. Attal M, Harousseau JL, Facon T, Guilhot F, Doyen C, Fuzibet JG, Monconduit M, Hulin C, Caillot D, Bouabdallah R et al (2003) Single versus double autologous stem-cell transplantation for multiple myeloma. N Engl J Med 349:2495–2502
3. Hernandez JM, Garcia-Sanz R, Golvano E, Blade J, Fernandez-Calvo J, Trujillo J, Soler JA, Gardella S, Carbonell F, Mateo G et al (2004) Randomized comparison of dexamethasone combined with melphalan versus melphalan with prednisone in the treatment of elderly patients with multiple myeloma. Br J Haematol 127:159–164
4. Richardson PG, Xie W, Mitsiades C, Chanan-Khan AA, Lonial S, Hassoun H, Avigan DE, Oaklander AL, Kuter DJ, Wen PY et al (2009) Single-agent bortezomib in previously untreated multiple myeloma: efficacy, characterization of peripheral neuropathy, and molecular correlations with response and neuropathy. J Clin Oncol 27:3518–3525
5. Dispenzieri A, Zhang L, Fonseca R (2006) Single agent bortezomib is associated with a high response rate in patients with high risk myeloma. A phase II study from the Eastern Cooperative Oncology Group. Blood 108:1006a
6. Harousseau JL, Mathiot C, Attal M (2008) Bortezomib/dexamethasone versus VAD as induction prior to autologous stem cell transplantation (ASCT) in previously untreated multiple myeloma (MM): updated data from IFM 2005/01 trial. J Clin Oncol 26:8505a
7. Sonneveld P, van der Holt B, Schmidt-Wolf IGH (2008) First analysis of HOVON-65/GMMG-HD4 randomized phase III trial comparing bortezomib, adriamycine, dexamethasone (PAD) vs VAD as induction treatment prior to high dose melphalan (HDM) in patients with newly diagnosed multiple myeloma (MM). Blood 112:653a
8. Cavo M, Patriarca F, Tacchetti P (2008) Superior complete response rate and progression-free survival after autologous transplantation with up-front Velcade-thalidomide-dexamethasone compared with thalidomide-dexamethasone in newly diagnosed multiple myeloma. Blood 112:158a
9. Rosinol L, Cibeira MT, Martinez J (2008) Thalidomide/dexamethasone (TD) vs. bortezomib (Velcade®)/thalidomide/dexamethasone (VTD) vs. VBMCP/VBAD/Velcade® as induction regimens prior to autologous stem cell transplantation (ASCT) in younger patients with multiple myeloma (MM): first results of a prospective phase III PETHEMA/Gem trial. Blood 112:654a
10. Jagannath S, Durie BG, Wolf JL, Camacho ES, Irwin D, Lutzky J, McKinley M, Potts P, Gabayan AE, Mazumder A et al (2009) Extended follow-up of a phase 2 trial of bortezomib alone and in combination with dexamethasone for the frontline treatment of multiple myeloma. Br J Haematol 146:619–626
11. Harousseau JL, Attal M, Leleu X, Troncy J, Pegourie B, Stoppa AM, Hulin C, Benboubker L, Fuzibet JG, Renaud M et al (2006) Bortezomib plus dexamethasone as induction treatment

prior to autologous stem cell transplantation in patients with newly diagnosed multiple myeloma: results of an IFM phase II study. Haematologica 91:1498–1505

12. Corso A, Barbarano L, Mangiacavalli S (2007) Bortezomib with HIG-dose dexamethasone as first line therapy in patients with multiple myeloma candidates to high-dose therapy. Blood 110:1051a
13. Na Nakorn T, Watanaboonyongcharoen P, Nipharak P (2007) Bortezomib with dexamethasone as induction therapy in newly diagnosed multiple myeloma: a preliminary study in Thai patients. Haematologica 92:181a
14. Rosinol L, Oriol A, Mateos MV, Sureda A, Garcia-Sanchez P, Gutierrez N, Alegre A, Lahuerta JJ, de la Rubia J, Herrero C et al (2007) Phase II PETHEMA trial of alternating bortezomib and dexamethasone as induction regimen before autologous stem-cell transplantation in younger patients with multiple myeloma: efficacy and clinical implications of tumor response kinetics. J Clin Oncol 25:4452–4458
15. Oakervee HE, Popat R, Curry N, Smith P, Morris C, Drake M, Agrawal S, Stec J, Schenkein D, Esseltine DL et al (2005) PAD combination therapy (PS-341/bortezomib, doxorubicin and dexamethasone) for previously untreated patients with multiple myeloma. Br J Haematol 129:755–762
16. Popat R, Oakervee HE, Hallam S, Curry N, Odeh L, Foot N, Esseltine DL, Drake M, Morris C, Cavenagh JD (2008) Bortezomib, doxorubicin and dexamethasone (PAD) front-line treatment of multiple myeloma: updated results after long-term follow-up. Br J Haematol 141:512–516
17. Jakubowiak A, Al-Zoubi A, Kendall T (2006) High rate of complete and near complete responses (CR/nCR) after initial therapy with bortezomib (Velcade), Doxil and dexamethasone (VDD) is further increased after autologous stem cell transplantation (ASCT). Blood 108:882a
18. Belch A, Reece D, Bahlis NJ (2007) Efficacy and safety of liposomal doxorubicin (Doxil/Caelyx), bortezomib (Velcade) and dexamethasone in the treatment of previously untreated multiple myeloma patients: impact of cytogenetic profile. Haematologica 92:180a
19. Palumbo A, Falco P, Gay F (2008) Bortezomib-doxorubicin-dexamethasone as induction prior to reduced intensity autologous transplantation followed by lenalidomide as consolidation/maintenance in elderly untreated myeloma patients. Blood 112:159a
20. Orlowski RZ, Peterson BL, Sanford B (2006) Bortezomib and pegylated liposomal doxorubicin as induction therapy for adult patients with symptomatic multiple myeloma: Cancer and Leukemia Group B Study 10301. Blood 108:239a
21. Kropff M, Liebisch P, Knop S, Weisel K, Wand H, Gann CN, Berdel WE, Einsele H (2009) DSMM XI study: dose definition for intravenous cyclophosphamide in combination with bortezomib/dexamethasone for remission induction in patients with newly diagnosed myeloma. Ann Hematol 88(11):1125–1130
22. Reeder CB, Reece DE, Kukreti V, Chen C, Trudel S, Hentz J, Noble B, Pirooz NA, Spong JE, Piza JG et al (2009) Cyclophosphamide, bortezomib and dexamethasone induction for newly diagnosed multiple myeloma: high response rates in a phase II clinical trial. Leukemia 23:1337–1341
23. Berenson JR, Yellin O, Woytowitz D, Flam MS, Cartmell A, Patel R, Duvivier H, Nassir Y, Eades B, Abaya CD et al (2009) Bortezomib, ascorbic acid and melphalan (BAM) therapy for patients with newly diagnosed multiple myeloma: an effective and well-tolerated frontline regimen. Eur J Haematol 82:433–439
24. Borrello I, Ferguson A, Huff CA (2006) Bortezomib and thalidomide treatment of newly diagnosed patients with multiple myeloma: efficacy and neurotoxicity. Blood 108:1006a
25. Chen S, Jiang B, Qiu L (2007) Bortezomib plus thalidomide treatment of newly diagnosed patients with multiple myeloma -efficacy and neurotoxicity: results of a multicenter study. Blood 110:3600a

26. Wang M, Giralt S, Delasalle K, Handy B, Alexanian R (2007) Bortezomib in combination with thalidomide-dexamethasone for previously untreated multiple myeloma. Hematology 12:235–239
27. Giralt S, Thandi R, Qazilbash M (2007) Retrospective comparison of transplant outcomes in patients with multiple myeloma according to induction therapy with thalidomide/dexamethasone (TD) with or without bortezomib (VTD). Blood 110:948a
28. Kaufman JL, Gleason C, Heffner LT (2007) Bortezomib, thalidomide, and dexamethasone as induction therapy for patients with symptomatic multiple myeloma. Blood 110:1055a
29. Badros A, Goloubeva O, Fenton R, Rapoport AP, Akpek G, Harris C, Ruehle K, Westphal S, Meisenberg B (2006) Phase I trial of first-line bortezomib/thalidomide plus chemotherapy for induction and stem cell mobilization in patients with multiple myeloma. Clin Lymphoma Myeloma 7:210–216
30. Barlogie B, Anaissie E, van Rhee F, Haessler J, Hollmig K, Pineda-Roman M, Cottler-Fox M, Mohiuddin A, Alsayed Y, Tricot G et al (2007) Incorporating bortezomib into upfront treatment for multiple myeloma: early results of total therapy 3. Br J Haematol 138:176–185
31. Barlogie B, Anaissie E, Shaughnessy J (2008) Ninety percent sustained complete response (CR) rate projected 4 years after onset of CR in gene expression profiling (GEP)-defined low-risk multiple myeloma (MM) treated with total therapy 3 (TT3): basis for GEP-risk-adapted TT4 and TT5. Blood 112:162a
32. Richardson P, Lonial S, Jakubowiak A (2008) Lenalidomide, bortezomib, and dexamethasone in patients with newly diagnosed multiple myeloma: encouraging efficacy in high risk groups with updated results of a phase I/II study. Blood 112:92a
33. Kumar S, Flinn IW, Noga SG (2008) Safety and efficacy of novel combination therapy with bortezomib, dexamethasone, cyclophosphamide, and lenalidomide in newly diagnosed multiple myeloma: initial results from the phase I/II multi-center EVOLUTION Study. Blood 112:93a
34. Mikala G (2008) Safe and effective mobilization of stem cells in multiple myeloma following priming by high-dose cyclophosphamide and bortezomib. Haematologica 95:655a
35. Roussel M, Huynh A, Moreau P (2008) Bortezomib (BOR) and high dose melphalan (HDM) as conditioning regimen before autologous stem cell transplantation (ASCT) for de novo multiple myeloma (MM): Final results of the IFM phase II study VEL/MEL. Blood 112:160a
36. Kaufman JL, Lonial S, Sinha R (2009) A randomized phase I study of melphalan and bortezomib for autologous transplant in myeloma. Clin Lymphoma Myeloma 9:364a
37. Ladetto M, Pagliano G, Ferrero S (2008) Major shrinking of residual tumor cell burden and achievement of molecular remissions in myeloma patients undergoing post-transplant consolidation with bortezomib, thalidomide and dexamethasone: A qualitative and quantitative PCR study. Blood 112:3683a
38. Mateos MV, Hernandez JM, Hernandez MT, Gutierrez NC, Palomera L, Fuertes M, az-Mediavilla J, Lahuerta JJ, de la Rubia J, Terol MJ et al (2006) Bortezomib plus melphalan and prednisone in elderly untreated patients with multiple myeloma: results of a multicenter phase 1/2 study. Blood 108:2165–2172
39. San Miguel JF, Schlag R, Khuageva NK, Dimopoulos MA, Shpilberg O, Kropff M, Spicka I, Petrucci MT, Palumbo A, Samoilova OS et al (2008) Bortezomib plus melphalan and prednisone for initial treatment of multiple myeloma. N Engl J Med 359:906–917
40. San-Miguel JF, Schlag R, Khuageva NK (2008) Updated follow-up and results of subsequent therapy in the phase III VISTA trial: bortezomib plus melphalan-prednisone versus melphalan-prednisone in newly diagnosed multiple myeloma. Blood 112:650a
41. Dimopoulos M, Richardson P, Schlag R (2009) Bortezomib-melphalan-prednisone is active and well tolerated in newly diagnosed multiple myeloma patients with moderately impaired renal function, and results in reversal of renal impairment: cohort analysis of the phase III VISTA study. J Clin Oncol 27(36):6086–6093

42. Mateos MV, Oriol A, Martinez J, Cibeira MT, de Paz R, Terol MJ, Garcia-Larana J, Bengoechea E, Martinez R, Martin A et al (2008) Bortezomib (Velcade)-melphalan-prednisone (VMP) versus Velcade- thalidomide-prednisone (VTP) in elderly untreated multiple myeloma patients: which is the best partner for Velcade: an alkylating or an immunomodulator agent? Blood 112:651a
43. Palumbo A, Bringhen S, Rossi G, Magaroto V, Di RF, Ria R, Offidani M, Nozzoli N, Patriarca F, Callea V et al (2008) A prospective, randomized, phase III study of bortezomib, melphalan, prednisone and thalidomide (VMPT) versus bortezomib, melphalan and prednisone (VMP) in elderly newly diagnosed myeloma patients. Blood 112:652a

The Use of Bortezomib in Autologous Transplantation for Multiple Myeloma

Amelia A. Langston and Sagar Lonial

Abstract High dose therapy and autologous peripheral blood or bone marrow transplantation represents a significant therapeutic advance for patients with multiple myeloma. While patients who undergo autologous transplantation have an improvement in overall survival compared to those who do not, the procedure is not curative. As the practice has become more widely used, it has become clear that methods to enhance the efficacy of transplantation are needed. The use of cytogenetics and FISH as a method by which to identify good or poor risk patients allows for more informed decisions about the durability of HDT and provides useful information to the clinician regarding the potential need of maintenance therapy in the post transplant period. Similarly, the use of more effective induction regimens have also had a significant impact on the depth of pre transplant response, as well as post transplant duration of remission. Finally, modifying the conditioning regimen itself represents yet another method by which the efficacy of autologous transplant can be further enhanced. These strategies to improve the outcomes of the transplant process for patients with multiple myeloma are currently the focus of existing and ongoing trials.

1 Introduction

Improvements in overall survival for patients with multiple myeloma were first achieved in the mid 1990s with widespread use of high dose therapy (HDT) and autologous transplantation [1, 2]. During this period, it became clear that HDT was able to overcome drug resistance, leading to clinical benefit even in the context of primary refractory disease [3]. The next steps have been focused on improving the

A.A. Langston and S. Lonial (✉)
Winship Cancer Institute, Emory University School of Medicine, 1365 Clifton Rd, Building C, Room 4004, Atlanta, GA 30322, USA
e-mail: sloni01@emory.edu

I.M. Ghobrial et al. (eds.), *Bortezomib in the Treatment of Multiple Myeloma*, Milestones in Drug Therapy,
DOI 10.1007/978-3-7643-8948-2_5, © Springer Basel AG 2011

efficacy of the transplant maneuver, which is critically important since not all patients gain the same magnitude of benefit from HDT. Optimization of the transplant process, particularly for those patients who benefit least from conventional HDT approaches, will be the next major advance toward improving outcomes.

Given the differential benefit of HDT among a heterogeneous group of patients, several approaches have emerged in an attempt to enhance the efficacy of HDT with the goal of improving progression free and overall survival. These approaches include risk-adapted patient selection, enhancement of the efficacy of initial induction regimens, refinements in the pre-transplant conditioning regimen, and the use of post-transplant maintenance therapy. Addressing each of these elements has the potential to improve the outcomes from conventional HDT and autologous transplant.

2 Enhancing Transplant Efficacy: Improving Patient Selection

It has been clear for over 20 years that all myeloma patients are not the same in terms of proliferative capacity and resistance to conventional agents. One of the oldest and most reliable methods for assessing risk among myeloma patients is the use of the beta-2 microglobulin [4, 5] (β2M), which is currently a major part of the International Staging System (ISS) [6]. It has been demonstrated that patients who present a higher β2M have both shorter duration of remission and shorter overall survival following HDT, compared with patients presenting a lower β2M. Unfortunately, the β2M analysis is dependant upon intrinsic renal function as well as other acute illness related factors that complicate its use as a gold standard for identifying patients with poor long term survival. The identification of patients with deletion of chromosome 13 by conventional karyotype analysis, and more recently [7], recognition that the presence of any karyotypic abnormality is associated with poor duration of remission and overall survival have further enhanced our ability to identify subsets of patients who may gain less benefit from HDT [8]. Fortunately, only about 15% of newly diagnosed patients have a karyotypic abnormality identified by conventional analysis, and thus, this poor risk group represents only a small subset of patients. However, we now know that among patients with a normal karyotype by conventional cytogenetic analysis, identification of t(4:14) or t(14:16) by FISH analysis is also associated with a poor prognosis [9]. The even less common 17p deletion rounds out the list of known poor risk criteria, all of which predict for a short duration of remission and short survival following HDT [10]. Following the identification of these high risk subsets of myeloma, there was a brief period when some experts discouraged the use of HDT in high risk patients although recent studies suggest that this approach may have been premature [11, 12]. Rather than discard HDT, current research focuses on methods to improve the efficacy of HDT for all patients in the era of modern anti-myeloma agents.

3 Enhancing Transplant Efficacy: The Role of Induction and Maintenance Therapy

In the induction setting, recent data suggest that more effective initial therapy may translate into better post-transplant outcomes. Trials from the IFM [13] and the GEIMEA [14] groups have demonstrated that the use of either bortezomib/dexamethasone (BD) or bortezomib, thalidomide and dexamethasone (VTD) are superior to the use of VAD or TD in terms of both pre-transplant and post-transplant response rates, as well as progression-free survival (PFS) following transplant. Whether this improved PFS can also be demonstrated among the patients with high risk disease, for whom we know that duration of remission following conventional transplant is short, remains to be seen.

The role of maintenance therapy has been controversial, particularly with respect to the question of whether PFS or overall survival should be the ultimate benchmark of success. Early trials using either steroids or interferon demonstrated modest benefit in small studies, but in larger studies have not demonstrated sufficient benefit to warrant use on a routine basis, particularly in the light of the significant side effects [15]. The use of thalidomide has been tested in a number of different trials with or without HDT, and although data are somewhat inconsistent [16], it appears that the use of thalidomide maintenance may offer benefit for patients who fail to achieve a CR following HDT [17, 18]. Among the patients who do achieve a CR following HDT, the benefit for maintenance thalidomide remains less clear. The use of bortezomib-based maintenance therapy is currently under study in a number of different large cooperative groups. In the Total Therapy-3 study, maintenance therapy with bortezomib plus thalidomide or lenalidomide in combination appeared to provide a significant survival benefit for high risk patients, suggesting that the use of maintenance therapy may be particularly important for maintaining remission in poor risk patients [19]. The use of bortezomib alone as maintenance is currently being evaluated in several European studies, and early data from the Hovon trial suggest that the use of bortezomib maintenance therapy may improve remission duration, although the impact on overall survival remains unclear.

4 Enhancing Transplant Efficacy: Improving Conditioning

In the final section, we will review attempts to improve post transplant outcomes by improving the transplant procedure itself. This is critically important, because while autologous transplant has been shown to improve overall survival, it is not a curative strategy. Eradication of resistant clones of plasma cells with achievement of lower levels of residual disease appears important for enhancing the efficacy of the transplant itself. The inherent sensitivity of extramedullary plasma cell collections to low doses of radiation therapy formed the basis for testing the combination of total body irradiation (TBI) plus melphalan as pre-transplant conditioning. Initial studies suggested significant activity for the combination, however, a subsequent randomized clinical trial from the IFM demonstrated that 200 mg/m^2 of melphalan was associated with better PFS and less toxicity than the combination of 140 mg/m^2 melphalan + TBI [20]. Several groups have also evaluated the use of combination chemotherapy conditioning regimens such as busulfan with cyclophosphamide (Bu/Cy), busulfan, cyclophosphamide and etoposide (Bu/Cy/VP-16) [18], or other combinations, however [21–23], none has proved superior to the use of MEL 200 in randomized clinical trials. There have been a few attempts to combine radioisotopes with cytotoxic chemotherapy, and while activity in early phase trials seemed encouraging, in the final analysis, they were often very toxic without clear cut benefits in terms of efficacy, so this approach has not been tested in phase III studies [24, 25]. Overall, combinations of poly-chemotherapy with or without radiotherapy have not appeared to enhance the efficacy of the transplant maneuver over what is seen with standard melphalan alone.

Several years ago, a series of laboratory observations suggested an inhibitory effect of proteasome inhibition on DNA repair enzymes. This work was predicated on the initial observation that the combination of alkylator agents and bortezomib resulted in synergistic plasma cell apoptosis, yet the mechanism was unclear [26]. In these experiments, it was clear that DNA PKCs were cleaved in a time- and dose-dependant fashion, consistent with impairment of alkylator induced DNA damage [27]. Based upon these observations, several groups have evaluated the feasibility and efficacy of combining high dose melphalan with various doses and schedules of bortezomib as conditioning prior to autologous transplantation, with the hope of enhancing the efficacy of the transplant maneuver.

The Arkansas team was one of the first to evaluate the combination of bortezomib and high dose melphalan prior to autologous transplant [28]. In this trial of 26 patients, the melphalan was given in multiple fractions with doses ranging from 150 to 250 mg/m^2, in conjunction with escalating doses of bortezomib and thalidomide. Toxicities were manageable, and 59% of patients achieved at least a nCR, with another 17% achieving a PR at 6 weeks post transplant. Based upon these data, the group is currently evaluating the combination of bortezomib, an immunomodulatory agent, dexamethasone and high dose melphalan in additional clinical studies.

The group from MDACC designed a phase II randomized trial testing a regimen they had previously used combining HD melphalan (HDM) with arsenic trioxide (ATO) and ascorbic acid (AA) as conditioning for autologous transplant [29]. In their previous experience with HDM+ATO+AA, the overall response rate for the trial was 85% with 25% of patients achieving a CR [30]. In the trial that incorporated bortezomib, 58 patients were randomized to receive either no bortezomib, three doses of bortezomib at 1 mg/m^2, or three doses of bortezomib at 1.5 mg/m^2 in conjunction with HDM+ATO+AA. In this randomized phase II experience, there was no difference in CR rate, PFS or OS between the three treatment arms, suggesting a lack of clinical benefit from the addition of bortezomib. It should be noted that several groups have now published clinical data suggesting that high doses of AA may have a negative effect on the efficacy of bortezomib [31]. This is further supported by laboratory work from Perrone et al. that suggests that even over the counter AA supplements may interefere with the binding of bortezomib to the proteasome [32], thus temporarily limiting its ability to block proteasome function [33]. Given the small size of the trial and the potential confounding effect of AA, the lack of benefit of bortezomib in the MDACC trial should be interpreted with caution.

The IFM has also evaluated the combination of HDM and bortezomib in a single arm study [34]. They used the standard HDM schedule, and added four doses of bortezomib 1 mg/m^2, two before and two after the melphalan. Among the 57 patients treated, nearly 70% of patients achieved at least a VGPR, and there was no apparent increase in hematologic or other toxicity.

At our center, we designed a randomized phase I/II trial to test the effect of sequence of administration of bortezomib in combination with HDM [35]. This was based on preclinical data suggesting that bortezomib given after melphalan is more effective at inducing plasma cell apoptosis than administering the bortezomib prior to melphalan. In order to test this hypothesis, we designed a randomized trial in which a single dose of bortezomib was given either before or after HDM, with a bortezomib dose escalation scheme that included 1.0–1.3 and 1.6 mg/m^2. We interrogated the marrow for the fraction of annexin V positive plasma cells prior to therapy, and on day 0 immediately before the transplant. In order to be eligible for the trial, patients had to have a measurable plasmacytosis in the marrow patients and could not have achieved a VGPR prior to transplant. A total of 39 patients were enrolled in the study, and the overall response rate was 95% with 21% of patients achieveing a CR and 55% achieving at least a VGPR. In contrast to previous trials using the combination of HDM and bortezomib, none of the patients in our trial had achieved a VGPR or better prior to transplant. Furthermore, the median number of prior lines of therapy for the group as a whole was two, with seven patients having received more than three prior lines of therapy. Toxicity of the combination was tolerable, and appeared similar to that observed with HDM alone. When sequence of administration was assessed using annexin V staining of the marrow plasma cells before and after therapy, it appeared that the sequence of bortezomib following HDM produced a greater degree of plasma cell apoptosis than did bortezomib given prior to HDM. The rate of achievement of at least a VGPR was also higher in the arm that received bortezomib after HDM, although the number of patients in each group was relatively small.

In conclusion, the use of HDT and autologous transplant is effective treatment for certain subsets of newly diagnosed myeloma patients, but improved approaches are needed, particularly for those patients with unfavorable biologic characteristics. We strive to improve outcomes for patients undergoing HDT through refinement in patient selection for conventional transplant, improvement in initial induction regimens, enhancement of pre-transplant conditioning regimens, and identification of patients who will benefit from post-transplant maintenance therapy. The use of bortezomib as part of the conditioning regimen is a particularly promising approach, with available data indicating better than expected response rates, even among patients who enter the transplant maneuver with advanced and/or poorly

Table 1 Features associated with prognosis for Myeloma

β2M	Elevated β2M is associated with higher ISS stage and poorer outcomes [4–6]
Cytogenetic abnormalities	Deletion 13 or any cytogenetic abnormality is associated with shorter duration of response following HDT [36, 37]
FISH abnormalities	t(4:14), Del 17p, and t(14:16) associated with shorter overall and PFS following HDT
	Del 13 alone by FISH does not have prognostic importance [10, 38, 39]
Elevated LDH	Elevated LDH is associated with more proliferation and poorer OS and PFS following HDT [40–43]
Circulating plasma cells	Increased number of circulating plasma cells, even if it does not meet criteria for plasma cell leukemia, is associated with poorer outcomes [44, 45]

Table 2 Methods to enhance the efficacy of HDT

Polychemotherapy	Addition of fludarabine, or busulfan based conditioning [18, 22, 46–48]
Addition of external beam radiation	TBI or hemi-body radiation [20, 46, 49]
Addition of radio-immuntherapy	Holmium DoTMP [24]
Tandem auto transplant	Role of tandem autologous transplants for all patients remains unclear [50–53]
Addition of novel agents	Arsenic trioxide, bortezomib [30, 34, 35]

responding disease. Obviously, these observations need to be further validated in large randomized clinical trials where the endpoints are not limited to response rate, but also progression free and overall survival Tables 1 and 2.

References

1. Attal M, Harousseau JL (1997) Standard therapy versus autologous transplantation in multiple myeloma. Hematol Oncol Clin North Am 11:133–146
2. Child JA, Morgan GJ, Davies FE et al (2003) High-dose chemotherapy with hematopoietic stem-cell rescue for multiple myeloma. N Engl J Med 348:1875–1883
3. Alexanian R, Weber D, Delasalle K, Handy B, Champlin R, Giralt S (2004) Clinical outcomes with intensive therapy for patients with primary resistant multiple myeloma. Bone Marrow Transplant 34:229–234
4. Greipp PR, Katzmann JA, O'Fallon WM, Kyle RA (1988) Value of beta 2-microglobulin level and plasma cell labeling indices as prognostic factors in patients with newly diagnosed myeloma. Blood 72:219–223
5. Greipp PR, Lust JA, O'Fallon WM, Katzmann JA, Witzig TE, Kyle RA (1993) Plasma cell labeling index and beta 2-microglobulin predict survival independent of thymidine kinase and C-reactive protein in multiple myeloma. Blood 81:3382–3387

6. Greipp PR, San Miguel J, Durie BG et al (2005) International staging system for multiple myeloma. J Clin Oncol 23:3412–3420
7. Shaughnessy J Jr, Tian E, Sawyer J et al (2003) Prognostic impact of cytogenetic and interphase fluorescence in situ hybridization-defined chromosome 13 deletion in multiple myeloma: early results of total therapy II. Br J Haematol 120:44–52
8. Shaughnessy J, Jacobson J, Sawyer J et al (2003) Continuous absence of metaphase-defined cytogenetic abnormalities, especially of chromosome 13 and hypodiploidy, ensures long-term survival in multiple myeloma treated with total therapy I: interpretation in the context of global gene expression. Blood 101:3849–3856
9. Avet-Loiseau H (2007) Role of genetics in prognostication in myeloma. Best Pract Res Clin Haematol 20:625–635
10. Avet-Loiseau H, Attal M, Moreau P et al (2007) Genetic abnormalities and survival in multiple myeloma: the experience of the Intergroupe Francophone du Myelome. Blood 109 (8):3489–3495
11. Dispenzieri A, Rajkumar SV, Gertz MA et al (2007) Treatment of newly diagnosed multiple myeloma based on mayo stratification of myeloma and risk-adapted therapy (mSMART): consensus statement. Mayo Clin Proc 82:323–341
12. Lonial S (2007) Designing risk-adapted therapy for multiple myeloma: the mayo perspective. Mayo Clin Proc 82:279–281
13. Harousseau JL, Mathiot C, Attal M, et al. (2008) Bortezomib/dexamethasone versus VAD as induction prior to autologous stem cell transplantion (ASCT) in previously untreated multiple myeloma (MM): Updated data from IFM 2005/01 trial. ASCO, Chicago, Ill: ASCO:8505
14. Cavo M, Tacchetti P, Patriarca F et al (2008) Superior complete response rate and progression-free survival after autologous transplantation with up-front velcade-thalidomide-dexamethasone compared with thalidomide-dexamethasone in newly diagnosed multiple myeloma. ASH Annual Meeting Abstracts 112:158.
15. Mihelic R, Kaufman JL, Lonial S (2007) Maintenance therapy in multiple myeloma. Leukemia 21:1150–1157
16. Barlogie B, Tricot G, Anaissie E et al (2006) Thalidomide and hematopoietic-cell transplantation for multiple myeloma. N Engl J Med 354:1021–1030
17. Attal M, Harousseau JL, Leyvraz S et al (2006) Maintenance therapy with thalidomide improves survival in patients with multiple myeloma. Blood 108:3289–3294
18. Brinker BT, Waller EK, Leong T et al (2006) Maintenance therapy with thalidomide improves overall survival after autologous hematopoietic progenitor cell transplantation for multiple myeloma. Cancer 106:2171–2180
19. Hoering A, Crowley J, Shaughnessy JD Jr et al (2009) Complete remission in multiple myeloma examined as time-dependent variable in terms of both onset and duration in total therapy protocols. Blood 114:1299–1305
20. Moreau P, Facon T, Attal M et al (2002) Comparison of 200 mg/m(2) melphalan and 8 Gy total body irradiation plus 140 mg/m(2) melphalan as conditioning regimens for peripheral blood stem cell transplantation in patients with newly diagnosed multiple myeloma: final analysis of the Intergroupe Francophone du Myelome 9502 randomized trial. Blood 99:731–735
21. Anagnostopoulos A, Aleman A, Ayers G et al (2004) Comparison of high-dose melphalan with a more intensive regimen of thiotepa, busulfan, and cyclophosphamide for patients with multiple myeloma. Cancer 100:2607–2612
22. Benson DM Jr, Elder PJ, Lin TS et al (2007) High-dose melphalan versus busulfan, cyclophosphamide, and etoposide as preparative regimens for autologous stem cell transplantation in patients with multiple myeloma. Leuk Res 31:1069–1075
23. Blanes M, de la Rubia J, Lahuerta JJ et al (2009) Single daily dose of intravenous busulfan and melphalan as a conditioning regimen for patients with multiple myeloma undergoing autologous stem cell transplantation: a phase II trial. Leuk Lymphoma 50:216–222

24. Christoforidou AV, Saliba RM, Williams P et al (2007) Results of a retrospective single institution analysis of targeted skeletal radiotherapy with (166)Holmium-DOTMP as conditioning regimen for autologous stem cell transplant for patients with multiple myeloma. Impact on transplant outcomes. Biol Blood Marrow Transplant 13:543–549
25. Giralt S, Bensinger W, Goodman M et al (2003) 166Ho-DOTMP plus melphalan followed by peripheral blood stem cell transplantation in patients with multiple myeloma: results of two phase 1/2 trials. Blood 102:2684–2691
26. Mitsiades N, Mitsiades CS, Richardson PG et al (2003) The proteasome inhibitor PS-341 potentiates sensitivity of multiple myeloma cells to conventional chemotherapeutic agents: therapeutic applications. Blood 101:2377–2380
27. Hideshima T, Hayashi T, Chauhan D, Akiyama M, Richardson P, Anderson K (2003) Biologic sequelae of c-Jun NH(2)-terminal kinase (JNK) activation in multiple myeloma cell lines. Oncogene 22:8797–8801
28. Pineda-Roman M, Fox MH, Hollmig KA et al (2006) Retrospective analysis of fractionated high-dose melphalan (F-MEL) and bortezomib-thalidomide-dexamethasone (VTD) with autotransplant (AT) support for advanced and refractory multiple myeloma (AR-MM). ASH Annual Meeting Abstracts. 108:3102
29. Qazilbash MH, Saliba RM, Pelosini M, et al. (2008) A randomized phase II trial of high-dose melphalan, ascorbic acid and arsenic trioxide with or without bortezomib in multiple myeloma. ASH Annual Meeting Abstracts. 112:3320
30. Qazilbash MH, Saliba RM, Nieto Y et al (2008) Arsenic trioxide with ascorbic acid and high-dose melphalan: results of a phase II randomized trial. Biol Blood Marrow Transplant 14:1401–1407
31. Zou W, Yue P, Lin N et al (2006) Vitamin C inactivates the proteasome inhibitor PS-341 in human cancer cells. Clin Cancer Res 12:273–280
32. Perrone G, Hideshima T, Ikeda H et al (2009) Ascorbic acid inhibits antitumor activity of bortezomib in vivo. Leukemia 23:1679–1686
33. Harvey RD, Nettles J, Wang B, Sun SY, Lonial S (2009) Commentary on Perrone et al.: 'vitamin C: not for breakfast anymore...if you have myeloma'. Leukemia 23:1939–1940
34. Roussel M, Moreau P, Huynh A et al (2010) Bortezomib and high dose melphalan as conditioning regimen before autologous stem cell transplantation in patients with de novo multiple myeloma: a phase II study of the Intergroupe Francophone du Myelome (IFM). Blood 115(1):32–7
35. Lonial S, Kaufman J, Torre C et al (2008) A randomized phase I trial of melphalan + bortezomib as conditioning for autologous transplant for myeloma: the effect of sequence of administration. ASH Annu Meet Abstr 112:3332
36. Fassas AB, Spencer T, Sawyer J et al (2002) Both hypodiploidy and deletion of chromosome 13 independently confer poor prognosis in multiple myeloma. Br J Haematol 118:1041–1047
37. Facon T, Avet-Loiseau H, Guillerm G et al (2001) Chromosome 13 abnormalities identified by FISH analysis and serum beta2-microglobulin produce a powerful myeloma staging system for patients receiving high-dose therapy. Blood 97:1566–1571
38. Rasmussen T, Knudsen LM, Dahl IM, Johnsen HE (2003) C-MAF oncogene dysregulation in multiple myeloma: frequency and biological relevance. Leuk Lymphoma 44:1761–1766
39. Bergsagel PL, Kuehl WM (2001) Chromosome translocations in multiple myeloma. Oncogene 20:5611–5622
40. Anagnostopoulos A, Gika D, Symeonidis A et al (2005) Multiple myeloma in elderly patients: prognostic factors and outcome. Eur J Haematol 75:370–375
41. Jurisic V, Colovic M (2002) Correlation of sera TNF-alpha with percentage of bone marrow plasma cells, LDH, beta2-microglobulin, and clinical stage in multiple myeloma. Med Oncol 19:133–139
42. Kurabayashi H, Kubota K, Tsuchiya J, Murakami H, Tamura J, Naruse T (1999) Prognostic value of morphological classifications and clinical variables in elderly and young patients with multiple myeloma. Ann Hematol 78:19–23

43. Suguro M, Kanda Y, Yamamoto R et al (2000) High serum lactate dehydrogenase level predicts short survival after vincristine-doxorubicin-dexamethasone (VAD) salvage for refractory multiple myeloma. Am J Hematol 65:132–135
44. Kumar S, Rajkumar SV, Kyle RA et al (2005) Prognostic value of circulating plasma cells in monoclonal gammopathy of undetermined significance. J Clin Oncol 23:5668–5674
45. Nowakowski GS, Witzig TE, Dingli D et al (2005) Circulating plasma cells detected by flow cytometry as a predictor of survival in 302 patients with newly diagnosed multiple myeloma. Blood 106:2276–2279
46. Desikan KR, Tricot G, Dhodapkar M et al (2000) Melphalan plus total body irradiation (MEL-TBI) or cyclophosphamide (MEL-CY) as a conditioning regimen with second autotransplant in responding patients with myeloma is inferior compared to historical controls receiving tandem transplants with melphalan alone. Bone Marrow Transplant 25:483–487
47. Talamo G, Claxton DF, Dougherty DW et al (2009) BU and CY as conditioning regimen for autologous transplant in patients with multiple myeloma. Bone Marrow Transplant 44: 157–161
48. Toor AA, Ayers J, Strupeck J et al (2004) Favourable results with a single autologous stem cell transplant following conditioning with busulphan and cyclophosphamide in patients with multiple myeloma. Br J Haematol 124:769–776
49. Wong JY, Rosenthal J, Liu A, Schultheiss T, Forman S, Somlo G (2009) Image-guided total-marrow irradiation using helical tomotherapy in patients with multiple myeloma and acute leukemia undergoing hematopoietic cell transplantation. Int J Radiat Oncol Biol Phys 73:273–279
50. Attal M, Harousseau JL, Facon T et al (2003) Single versus double autologous stem-cell transplantation for multiple myeloma. N Engl J Med 349:2495–2502
51. Moreau P, Hullin C, Garban F et al (2006) Tandem autologous stem cell transplantation in high-risk de novo multiple myeloma: final results of the prospective and randomized IFM 99-04 protocol. Blood 107:397–403
52. Kumar A, Kharfan-Dabaja MA, Glasmacher A, Djulbegovic B (2009) Tandem versus single autologous hematopoietic cell transplantation for the treatment of multiple myeloma: a systematic review and meta-analysis. J Natl Cancer Inst 101:100–106
53. Cavo M, Tosi P, Zamagni E et al (2007) Prospective, randomized study of single compared with double autologous stem-cell transplantation for multiple myeloma: Bologna 96 clinical study. J Clin Oncol 25:2434–2441

Bortezomib in Relapsed and Relapsed/ Refractory Multiple Myeloma

Jatin J. Shah and Robert Z. Orlowski

Abstract Proteasome inhibition was validated as a rational therapeutic approach when bortezomib (VELCADE®), the first-in-class proteasome inhibitor, was approved by the Food and Drug Administration in 2003. This initial approval, which was for multiple myeloma patients who had received at least two prior therapies, and whose disease had demonstrated progression on the last of these, also made bortezomib the first new agent to be available for myeloma in over a decade. Since that time, further studies of bortezomib both alone, and as part of rationally designed, molecularly based combination regimens, have shown the versatility of this agent, and made it a universally accepted standard of care. Importantly, this use of bortezomib has been associated with ever increasing overall response rates and response qualities, and, most importantly, long-term outcome measures, including overall survival. These findings have rightfully entrenched bortezomib as one of the most important parts of our chemotherapeutic armamentarium against multiple myeloma. In this chapter, we will review the role of bortezomib as a single-agent, and in combination with other chemotherapeutics, to combat myeloma in the relapsed and relapsed/refractory settings. Also, emerging data about retreatment with bortezomib will be presented to provide insights into the possible role of this agent in the modern era, in which patients may have already been previously treated as part of induction therapy with bortezomib. Finally, approaches that show promise to overcome primary or secondary resistance to

J.J. Shah
Department of Lymphoma & Myeloma, Division of Cancer Medicine, The University of Texas M. D. Anderson Cancer Center, Houston, TX, USA

R.Z. Orlowski (✉)
Department of Lymphoma & Myeloma, Division of Cancer Medicine, The University of Texas M. D. Anderson Cancer Center, Houston, TX, USA
Department of Experimental Therapeutics, Division of Cancer Medicine, The University of Texas M. D. Anderson Cancer Center, 1515 Holcombe Blvd., Unit 429, Houston, TX 77030-4009, USA
e-mail: rorlowsk@mdanderson.org

I.M. Ghobrial et al. (eds.), *Bortezomib in the Treatment of Multiple Myeloma*, Milestones in Drug Therapy,
DOI 10.1007/978-3-7643-8948-2_6,

bortezomib will be examined, to determine if this agent may be of benefit in a number of myeloma settings in each individual patient.

1 Single-Agent Bortezomib in the Relapsed/Refractory Setting

The clinical anti-myeloma activity of bortezomib was first described in the initial phase I trial of this agent, then known as PS-341, which evaluated patients with relapsed and/or refractory hematologic malignancies. In this study, a clinical benefit was reported in each of the nine patients with plasma cell dyscrasias, including one with a durable complete remission (CR), who also was the first myeloma patient ever to be treated with bortezomib [1]. This activity against relapsed/refractory myeloma was then further evaluated in two multi-center phase II trials, the Study of Uncontrolled Multiple Myeloma Managed with Proteasome Inhibition Therapy (SUMMIT)[2, 3], and the Clinical Response and Efficacy Study of Bortezomib in the Treatment of Relapsing Multiple Myeloma (CREST)[4, 5]. Both administered bortezomib as an intravenous push on days 1, 4, 8, and 11 of every 3-week cycle, which remains to this day the most commonly used schedule. Patients on the SUMMIT trial received an initial dose of 1.3 mg/m^2, while the smaller CREST study also explored a lower dose of 1.0 mg/m^2. On the SUMMIT trial, the overwhelming majority of patients had disease that was refractory to their last therapy, and a partial response (PR) or better was seen in 27% of the 193 evaluable patients. Among these were 10% who achieved a CR or near-CR (nCR), and for the population as a whole, the median time to progression (TTP) was 7 months, which more than doubled the median TTP for patients on their previous line of therapy. Adverse events of note included thrombocytopenia (in 28% of patients), fatigue (12%), peripheral neuropathy (12%), and neutropenia (11%). Thrombocytopenia [6] and neuropathy [7–9] induced by bortezomib have been studied extensively, and will in part be reviewed in a subsequent chapter; these toxicities have been found to be predictable, transient, and reversible.

Patients with relapsed or refractory disease after front-line therapy treated on the CREST study with bortezomib either alone at 1.3 mg/m^2, or with the addition of dexamethasone, had a 50% response rate [4, 5]. In the cohort that received 1.0 mg/m^2, the response rate was 38%, which was balanced by a lower likelihood of developing some adverse events such as diarrhea, vomiting, neuropathy, and dyspnea.

Data from the SUMMIT study supported accelerated approval of bortezomib for relapsed/refractory myeloma, and led to a randomized phase III trial, the Assessment of Proteasome Inhibition for Extending Remissions, or APEX study. This trial targeted patients with disease that had relapsed after one to three prior lines of therapy [10], who were randomized to receive either single-agent bortezomib or oral dexamethasone. The most common adverse events of all grades in the bortezomib arm included diarrhea (57% of patients), nausea (57%), fatigue (42%), constipation (42%), neuropathy (36%), vomiting (35%), anorexia (35%), and thrombocytopenia (35%). Notably, the majority of these were of grade 1 or

2 severity, and grade 4, as well as serious adverse events, were comparable on the two arms. At the initial report of the data, at least a PR was seen in 38% of bortezomib-treated patients, with 9% CR, compared to 18% and less than 1%, respectively, for dexamethasone. Continued therapy led to an improvement in the bortezomib response rate to 43%, while patients on the dexamethasone arm saw no such added benefit. Long-term outcomes were superior on the bortezomib arm as well, with a TTP of 6.22 months compared to only 3.49 months with dexamethasone, while median survival was 29.8 months compared to 23.7 months, respectively. These data supported full approval of bortezomib for patients with relapsed and/or refractory myeloma who had received at least one prior therapy.

A number of subanalyses of the data from these trials yielded important information that has helped to shape our current use of bortezomib both in the relapsed and front-line settings. This agent showed an excellent safety and efficacy profile in patients with high-risk disease defined by clinical criteria such as advanced age and elevation of the β_2-microglobulin [11, 12]. Along a similar vein, bortezomib and bortezomib-based regimens were safely administered in the setting of renal impairment, and provided equivalent to perhaps even greater response rates [13–16], with reversal of renal failure in approximately 30–40%.

2 Combination Regimens Incorporating Bortezomib

Modulation of proteasome function can be used not just as an approach by itself, but has been validated pre-clinically as a means to enhance sensitivity to other agents and, in many cases, to overcome drug resistance, as detailed in several excellent reviews [17, 18], and in an earlier chapter. Corticosteroids form the backbone of many anti-myeloma regimens, and the combination of bortezomib and dexamethasone was evaluated in the SUMMIT and CREST studies based on pre-clinical data showing the potential of at least additive to possibly synergistic activity [19]. Both allowed addition of dexamethasone at 20 mg on the day of each dose of bortezomib, and the day after, in patients who had progressive disease after two cycles, or stable disease after the first four cycles. With this approach, improvements in the response quality were seen in up to one-third of such patients, and other trials have shown that adding corticosteroids can improve response rates to 60% or more [2, 20, 21], without comparable increases in toxicity. Notably, patients with disease that had previously been steroid-refractory were able to respond to regimens of bortezomib and dexamethasone, suggesting that bortezomib overcomes drug resistance in vivo as well.

A second two-drug combination for which there is extensive support is the regimen of bortezomib with pegylated liposomal doxorubicin (Doxil®; PLD), which was designed based on the ability of PLD to suppress the anti-apoptotic induction by proteasome inhibitors of mitogen-activated protein kinase phosphatase [22, 23]. The initial phase I study targeting patients with all hematologic malignancies showed it to be a tolerable regimen [24], and activity was seen against multiple myeloma, as well as acute myeloid leukemia and non-Hodgkin lymphoma.

With regard to myeloma, a 73% response rate was seen among 22 evaluable patients, which was evenly divided between patients achieving a PR or CR/nCR. Long-term outcomes among these evaluable patients were favorable as well, with a TTP of 9.3 months, and a time to retreatment of 24 months [25], which provided the impetus for a subsequent randomized, international study [26]. In that trial, addition of PLD increased the overall response rate modestly, but did improve the proportion of patients in either a CR or very good PR (VGPR) by 42%. Median TTP was the primary endpoint and, interestingly, was identical with bortezomib/PLD on this study to that seen in the phase I trial at 9.3 months, and significantly better than that for bortezomib alone at 6.5 months ($P = 0.000004$). This enhanced response quality also translated into significant prolongations in the median duration of response and progression-free survival (PFS), and an early trend towards an improved overall survival (OS). Further supporting the ability of bortezomib to overcome prior drug resistance, bortezomib/PLD was more active than bortezomib alone in patients whose disease had been exposed to other anti-myeloma agents, including anthracyclines and immunomodulatory drugs (IMiDs), and was safe and effective in patients with renal compromise [27, 28]. Further studies have added other active anti-myeloma agents to the bortezomib/PLD backbone, such as thalidomide or thalidomide and dexamethasone. These combinations have yielded overall response rates of 63–81%, and CR/nCR rates of 17–33% (Table 1), supporting their use in this setting.

Alkylating agents represent another class of drugs with which bortezomib has been combined, including cyclophosphamide, melphalan, and bendamustine. These have been rationally designed based on findings that proteasome inhibition suppresses DNA damage repair pathways, supporting combinations with DNA damaging agents [29, 30]. Regimens of cyclophosphamide and bortezomib have been evaluated both in the up-front and relapsed settings, with studies in the latter showing response rates of 68–82%, and CR/nCR rates of 16–32% (Table 2). Melphalan-based combinations with bortezomib, ranging from the simple doublet to four-drug programs, such as bortezomib, melphalan, prednisone, and thalidomide,

Table 1 Bortezomib-based combinations with anthracyclines in the relapsed and relapsed and/or refractory setting

Regimen	Study phase	ORR/CR+nCR	TTP	Reference
Bortezomib + PLD	I	73%/36% (*n* = 22)	9.3 months	[24, 25]
Bortezomib + PLD vs.	III	44%/13% (*n* = 324)	9.3 months	[26]
Bortezomib		41%/10% (*n* = 322)	6.5 months	
Bortezomib + PLD + thalidomide	II	56%/22% (*n* = 23)	10.9 months[a]	[50]
Bortezomib + PLD + thalidomide + dexamethasone	II	81%/52% (*n* = 42)	19 months	[56]
Bortezomib + doxorubicin + dexamethasone	II	67%/9% (CR) (*n* = 64)	N/A	[57]

CR complete response, *n* cohort size, *nCR* near-complete response, *N/A* not available, *ORR* overall response rate (defined as a partial response or better), *PLD* pegylated liposomal doxorubicin, *TTP* time to progression

[a]Progression-free survival

Table 2 Bortezomib-based combinations with alkylating agents

Regimen	Study phase	ORR/CR+nCR	Selected grade 3/4 adverse events	PFS	Reference
Bortezomib + bendamustine + dexamethasone	I	57%/0% ($n = 7$)	Thrombocytopenia 29%, neutropenia 14%, anemia 29%, infection 29%	8 mos.[a]	[58]
Bortezomib + melphalan	I/II	50%/15% ($n = 46$)	Thrombocytopenia 27%, neutropenia 31%, anemia 13%, neuropathy 8%	9 mos.	[59, 60]
Bortezomib + melphalan +/− dexamethasone	I/II	68%/23% ($n = 55$)	Thrombocytopenia 62%, neutropenia 57%, infection 21%, neuropathy 15%	10 mos.	[61]
Bortezomib + melphalan + thalidomide +/− dexamethasone	II	66%/13% ($n = 62$)	Thrombocytopenia 23%, neutropenia 10%, infection 15%, neuropathy 15%	9.3 mos.[a]	[62]
Bortezomib + melphalan + thalidomide + prednisone	I/II	67%/17% ($n = 30$)	Thrombocytopenia 33%, neutropenia 43%, anemia 13%, neuropathy 7%	61% at 1 year	[51]
Bortezomib + cyclophosphamide + prednisone	I/II	68%/32% ($n = 37$)	Cycle 1 only: Nausea 8%, thrombocytopenia 13%, neutropenia 8%	15 mos.	[49]
Bortezomib + cyclophosphamide + dexamethasone	Retrospective	75%/31% ($n = 16$)	Thrombocytopenia 44%, neutropenia 19%, infection 13%, neuropathy 12%	7 mos.	[63]
Bortezomib + cyclophosphamide + dexamethasone	II	82%/16% ($n = 50$)	Thrombocytopenia 53%, leukopenia 20%, infection 23%, neuropathy 21%	12 mos.	[64]

CR complete response, *mos.* months, *n* cohort size, *nCR* near-complete response, *ORR* overall response rate (defined as at least a partial response), *PFS* progression-free survival, *Ref.* reference

[a]Time to progression

have provided response rates of up to 50–68% (Table 2). Most recently, bendamustine has been added to bortezomib, and in all of these cases an acceptable safety profile has been described. However, as could be predicted, several of these phase I/II studies have shown what would appear to be an increased risk of cytopenias and infectious complications compared to what might be expected for bortezomib alone. Larger studies in a randomized fashion will be needed to better evaluate the most appropriate use of these regimens against myeloma in the relapsed and relapsed/refractory settings.

The fourth major class of agents combined with bortezomib have been the IMiDs, including thalidomide or lenalidomide. Addition of thalidomide and dexamethasone to bortezomib has yielded overall response rates (ORR) of approximately 50–60% (Table 3), while the incorporation of lenalidomide and dexamethasone provided a 67% ORR, with responses reported to be independent of adverse cytogenetics [31]. Notably, this regimen showed activity even in patients who had had disease previously progress through bortezomib with dexamethasone, or lenalidomide with dexamethasone, highlighting one strategy for such patients in the modern era, namely to mix and match drugs that were used in previous lines of therapy into new regimens.

Other bortezomib-based combinations that have been studied for relapsed and/or refractory myeloma have added novel agents such as histone deacetylase (HDAC) inhibitors, protein kinase B/Akt inhibitors, heat shock protein (HSP)-90 inhibitors, and monoclonal antibodies (Table 4). Many of these have been rationally designed based on findings that proteasome inhibition induces compensatory stress responses, which includes up-regulation of transcripts of the ubiquitin-proteasome pathway components [32]. HDAC inhibition suppresses the expression of proteasome subunits and other ubiquitin proteasome pathway members [33], supporting the combination of bortezomib and an HDAC inhibitor. Bortezomib also induces aggresome formation as an alternative means to remove misfolded proteins [34, 35], supporting the use of HDAC and HSP-90 inhibitors, which block this pathway [34–36]. This preclinical work has translated into two phase I clinical trials to evaluate the safety and efficacy of the combination of vorinostat and bortezomib [37–39]. The first study led to an ORR of 50% with stable disease or better in 13 of 16 patients, and similar results from the second study, where stable disease or better

Table 3 Bortezomib-based combinations with immunomodulatory agents

Regimen	Study phase	ORR/CR+nCR	PFS	Reference
Bortezomib + thalidomide + dexamethasone	II	47%/12% (n = 18)	N/A	[56]
Bortezomib + thalidomide + dexamethasone	I/II	63%/24% (n = 85)	6 mos.	[65]
Bortezomib + lenalidomide + dexamethasone	I/II	67%/24% (n = 64)	N/A	[31]

CR complete response, *mos.* months, *n* cohort size, *N/A* not available, *nCR* near-complete response, *ORR* overall response rate (defined as at least a partial response), *PFS* progression-free survival

Table 4 Other bortezomib-based combination regimens

Regimen	Study phase	ORR	Reference
Bortezomib + samarium lexidronam	I	13% ($n = 24$)	[66]
Bortezomib + perifosine	I/II	16% ($n = 57$)	[67]
Bortezomib + tanespimycin	I/II	46% ($n = 40$)	[68]
Bortezomib + vorinostat	I	26% ($n = 34$)	[39]
Bortezomib + panobinostat	I	36%[a] ($n = 14$)	[69]
Bortezomib + siltuximab (CNTO 328)	II	57% ($n = 21$)	[70]
Bortezomib + ascorbic acid, arsenic trioxide	I	9% ($n = 22$)	[71]
Bortezomib + vorinostat, dexamethasone	I	43% ($n = 21$)	[39]
Bortezomib + temsirolimus (CCI-779)	I	7% ($n = 15$)	[72]

n cohort size, *ORR* overall response rate (defined as at least a partial response)
[a]Three of five responders had dexamethasone added with the second cycle.

was seen in all 17 patients, including many patients with previous bortezomib refractory disease. Inhibition of protein kinase B/Akt enhances the activity of bortezomib by suppressing anti-apoptotic Akt activity, which is induced by bortezomib [40], and encouraging data in combination with perifosine have been obtained (Table 4). Similarly, HSP-90 inhibitors, in addition to reducing aggresome formation, also suppress pro-survival signaling through this chaperone [41], and the bortezomib/tanespimycin combination has shown interesting activity (Table 4). Importantly, several of these regimens have shown the ability to induce responses in some patients whose disease was previously refractory to bortezomib. Finally, blockade of interleukin (IL)-6 signaling with siltuximab (CNTO 328) is an interesting approach, since it reduces bortezomib-mediated induction of HSP-70 and of myeloid cell leukemia (Mcl)-1 [42], and has shown promise clinically (Table 4). A randomized clinical trial of siltuximab and bortezomib compared to bortezomib alone has already been completed and the data are eagerly awaited, while phase III studies of these other combinations are either underway or planned.

3 Bortezomib and High-Risk Cytogenetic Features

The most common unfavorable cytogenetic abnormalities detected by fluorescence in situ hybridization in patients with multiple myeloma include del(13q), t(4;14), and del(17p13). These have historically been associated with a more aggressive disease phenotype, and inferior long-term outcome measures after standard- or high-dose therapy approaches. Notably, bortezomib based therapy has resulted in promising results that support the possibility that this agent can overcome at least some of these high-risk features (Table 5). Analyses from the SUMMIT study demonstrated no difference in response rates or survival between patients with or without del(13) by metaphase cytogenetics [11, 43]. In the larger APEX study, patients with del(13) in the dexamethasone arm had a lower ORR and OS compared to patients without del(13), while patients with del(13) treated with bortezomib had a similar ORR and OS as those patients without del(13).

Table 5 Bortezomib in patients with high risk cytogenetics in the relapsed/refractory setting

Regimen	Study type	Chromosomes evaluated	Difference in ORR	Difference in PFS/OS	Reference
Bortezomib −/+ dexamethasone	Retrospective (APEX/ SUMMIT)	Del(13)	No	No	[43]
Bortezomib + dexamethasone	Retrospective	Del(13), t(4;14), t(11;14), Amp *CKS1B*	No	No	[46]
Bortezomib −/+ dexamethasone	Prospective	Del(13), 14q32 translocation	No	Yes, shorter PFS/OS in del (13) only	[45]
Bortezomib + lenalidomide + dexamethasone	Prospective	Abnormal cytogenetics by metaphase, del(13) by FISH	No	N/A	[31]

Amp amplification, *APEX* Assessment of Proteasome Inhibition for Extending Remissions, *CKS1* Cdc kinase subunit 1, *Del* deletion, *FISH* fluorescence in situ hybridization, *nCR* near-complete response, *ORR* overall response rate (defined as at least a partial response), *OS* overall survival, *PFS* progression-free survival, *SUMMIT* Study of Uncontrolled Multiple Myeloma Managed with Proteasome Inhibition Therapy

Additional analyses have also suggested that bortezomib was able to overcome adverse cytogenetic prognostic factors in the relapsed or relapsed/refractory setting, such as del(13) and t(4;14), especially with regard to ORR [44–46]. Whether this effect will translate into an impact on PFS and OS compared to standard therapies is less clear at this time, since these studies have been of limited sample size and follow-up duration, and were not performed in a prospective fashion with stratification for risk factors.

4 Alternate Dosing Schedules Utilizing Bortezomib

Bortezomib was originally approved in the relapsed/refractory setting on a day 1, 4, 8, and 11 administration schedule of every 21-day cycle, but modifications of this regimen have been studied with the goal of enhancing convenience, and/or reducing drug-related toxicities. This was first done in the APEX study, where patients were initially treated with eight cycles of the above regimen, and could then continue therapy with day 1, 8, 15, and 22 dosing of every 35-day cycle for three cycles. While this therapy proved to be tolerable, the benefits of such reduced-intensity dosing were not established, since there was no control group containing patients randomized to observation, or standard bortezomib dosing. Another approach taken by some investigators is to use weekly bortezomib, but at a higher dose of 1.6 mg/mg^2, such as in the phase II study by Hainsworth et al. of 40 patients with one or two previous therapies who received bortezomib for four consecutive weeks of every 5 week cycle [47]. An ORR of 55% was seen, along with a duration of response of

16 months, and an encouraging PFS of 9.6 months. Though this schedule was also tolerable, with some grade 3/4 toxicities seemingly occurring at a lower rate than would be expected from the APEX data, including thrombocytopenia (20%), other toxicities were comparable or even increased in frequency, including neutropenia (13%), fatigue (15%), diarrhea (13%), and neuropathy (10%).

Hearteningly, other studies, predominantly using bortezomib in combination with other agents, have suggested that a reduced-frequency dosing schedule may have both enhanced efficacy and tolerability. Suvannasankha et al. prospectively evaluated the two-drug combination of bortezomib and methylprednisolone, with each agent being given weekly for 3 weeks followed by 1 week off, and among 18 patients the ORR was 62%, with a TTP of 6.6 months and OS of 20.2 months [48]. Toxicity was minimal, with grade 3 events limited to two patients with neuropathy, and one each with congestive heart failure and a gastrointestinal event. Similarly, Reece et al. evaluated bortezomib with cyclophosphamide and prednisone using both a standard bortezomib and weekly dosing schema, with the latter being at 1.5 mg/m^2 [49]. This weekly bortezomib schedule appeared to yield similar results to those with standard dosing, with an ORR of 92% and 100%, respectively, and while grade 3/4 thrombocytopenia was seen in 43% of cycles with standard dosing, no such episodes, nor any of neuropathy, were seen with the weekly schedule. Using a steroid-sparing regimen of three drugs, Chanan-Khan et al. evaluated bortezomib with PLD and thalidomide, with bortezomib administered on days 1, 4, 15 and 18, while PLD at 20 mg/m^2 was given on days 1 and 15 every 28 days [50]. Despite the fact that the majority of patients had refractory disease, the ORR remained significant at 55%, and only one patient out of 23 experienced grade 3 or worse neuropathy. Finally, four-agent combinations have been evaluated as well, such as bortezomib, melphalan, prednisone, and thalidomide by Palumbo et al. [51], with bortezomib administered at 1.0 mg/m^2, 1.3 mg/m^2, or 1.6 mg/m^2, all on days 1, 4, 15, and 22 every 5 weeks. Thirty patients with relapsed or refractory disease were enrolled, and 67% achieved at least a PR, with only two patients experiencing grade 3 neuropathy, and no grade 4 toxicities were recorded.

Taken together, a fairly consistent pattern with alternative dosing schedules has emerged and seems to indicate that when bortezomib is combined with other drugs, it can be given with reduced frequency and is generally associated with an improved toxicity profile, including reductions in cytopenias and peripheral neuropathy. Robust response rates have been reported, and limited longer-term follow-up data suggest that these responses will be durable, but larger randomized trials will be needed to validate this hypothesis further.

5 Retreatment with Bortezomib

Prior to the era of novel agents, it was rare for patients to achieve complete remissions in the relapsed and/or refractory setting outside of the use of high dose chemotherapy with autologous stem cell rescue. The ability of bortezomib to induce such high quality responses led to interest in the possibility of reusing this proteasome inhibitor in patients who had previously obtained a good benefit. Several studies have been reported that support the viability of this approach targeting patients with previous bortezomib exposure, which typically has been at least 60 days prior to the start of retreatment. In one case series of a cohort of 22 patients previously treated on the SUMMIT, CREST, or APEX studies who were later rechallenged off protocol, a 50% retreatment response rate was reported with bortezomib either alone or in combination with another agent, compared to an initial 68% response rate [52]. A second retrospective study of 82 patients who had experienced an initial response rate of 59% was less encouraging, with a retreatment response rate of 21% overall [53], though in the subgroup who had previously achieved a VGPR, the ORR at rechallenge was 44%. Finally, a third study of sixty patients, all of whom had responded to bortezomib previously, noted a 63% response rate when bortezomib or bortezomib and dexamethasone was reapplied [54]. Interestingly, patients with a treatment-free interval of 6 or more months had a higher clinical benefit ratio, with almost 90% having at least stable disease, while only slightly more than 60% of those with a shorter treatment-free interval benefited in this fashion. Patients with less sensitive disease may still benefit from retreatment with bortezomib, especially if additional agents are combined with this proteasome inhibitor if it by itself does not initially result in a response. Using this approach, with the addition of dexamethasone, or melphalan and prednisone after four cycles if no response was seen for single-agent therapy, a response rate of 73% was achieved [55].

6 Conclusions

Bortezomib was initially approved for patients with relapsed/refractory myeloma, and it remains a standard of care in this setting, where it is our most active drug as a single agent. Its unique benefits include an excellent safety and efficacy profile in all patients, including those with high-risk disease based on clinical criteria such as renal failure and older age, as well as high-risk disease by cytogenetic criteria such as del(13). Moreover, when PLD is added to bortezomib, virtually all clinically relevant patient subgroups, with the exception of those with a low β2-microglobulin and del(13), derive an enhanced benefit from this rationally designed regimen, compared to bortezomib alone. Other drugs have been successfully added to bortezomib as well, including standard cytotoxic agents such as cyclophosphamide and bendamustine, immunomodulatory drugs, and novel, targeted therapeutics such as perifosine, tanespimycin, vorinostat, and siltuximab, often using different dosing schedules. Many of these have shown evidence of enhanced response rates, often with no significant increase in toxicity, and indeed in some cases a decrease in side effects, such as neuropathy, has been seen. Data from active and pending randomized studies, which will be presented in the coming years, will be helpful in determining the role of these regimens in our chemotherapeutic armamentarium against myeloma. Excitingly, several of these regimens have shown the ability to overcome prior clinical bortezomib resistance, suggesting the possibility that bortezomib will be useful in several lines of therapy for myeloma. This may especially be true for patients who had high quality responses previously, and if bortezomib is given in different combinations in the retreatment setting than were used previously. It is now incumbent upon the myeloma research community to more clearly define the mechanisms of resistance that are activated by prior exposure to bortezomib, and to develop assays that would allow classification of each patient's disease based upon which mechanism is most prominent. With this information in hand, we will then be able to individualize therapy by applying the regimen that is most likely to overcome resistance in each patient, thereby optimizing the benefits of this drug, which has proven itself so uniquely suited to the treatment of multiple myeloma.

Acknowledgements J.J.S. would like to acknowledge support from the National Cancer Institute (P01 CA124787). R.Z.O., a Leukemia & Lymphoma Society Scholar in Clinical Research, and would also like to acknowledge support from the National Cancer Institute (R01 CA102278 and P01 CA124787).

References

1. Orlowski RZ, Stinchcombe TE, Mitchell BS, Shea TC, Baldwin AS, Stahl S, Adams J, Esseltine DL, Elliott PJ, Pien CS et al (2002) Phase I trial of the proteasome inhibitor PS-341 in patients with refractory hematologic malignancies. J Clin Oncol 20:4420–4427
2. Richardson PG, Barlogie B, Berenson J, Singhal S, Jagannath S, Irwin D, Rajkumar SV, Srkalovic G, Alsina M, Alexanian R et al (2003) A phase 2 study of bortezomib in relapsed, refractory myeloma. N Engl J Med 348:2609–2617
3. Richardson PG, Barlogie B, Berenson J, Singhal S, Jagannath S, Irwin DH, Rajkumar SV, Srkalovic G, Alsina M, Anderson KC (2006) Extended follow-up of a phase II trial in relapsed, refractory multiple myeloma: final time-to-event results from the SUMMIT trial. Cancer 106:1316–1319
4. Jagannath S, Barlogie B, Berenson J, Siegel D, Irwin D, Richardson PG, Niesvizky R, Alexanian R, Limentani SA, Alsina M et al (2004) A phase 2 study of two doses of bortezomib in relapsed or refractory myeloma. Br J Haematol 127:165–172
5. Jagannath S, Barlogie B, Berenson JR, Siegel DS, Irwin D, Richardson PG, Niesvizky R, Alexanian R, Limentani SA, Alsina M et al (2008) Updated survival analyses after prolonged follow-up of the phase 2, multicenter CREST study of bortezomib in relapsed or refractory multiple myeloma. Br J Haematol 143:537–540
6. Lonial S, Waller EK, Richardson PG, Jagannath S, Orlowski RZ, Giver CR, Jaye DL, Francis D, Giusti S, Torre C et al (2005) Risk factors and kinetics of thrombocytopenia associated with bortezomib for relapsed, refractory multiple myeloma. Blood 106:3777–3784
7. Richardson PG, Briemberg H, Jagannath S, Wen PY, Barlogie B, Berenson J, Singhal S, Siegel DS, Irwin D, Schuster M et al (2006) Frequency, characteristics, and reversibility of peripheral neuropathy during treatment of advanced multiple myeloma with bortezomib. J Clin Oncol 24:3113–3120
8. Badros A, Goloubeva O, Dalal JS, Can I, Thompson J, Rapoport AP, Heyman M, Akpek G, Fenton RG (2007) Neurotoxicity of bortezomib therapy in multiple myeloma: a single-center experience and review of the literature. Cancer 110:1042–1049

9. Richardson PG, Sonneveld P, Schuster MW, Stadtmauer EA, Facon T, Harousseau JL, Ben-Yehuda D, Lonial S, Goldschmidt H, Reece D et al (2009) Reversibility of symptomatic peripheral neuropathy with bortezomib in the phase III APEX trial in relapsed multiple myeloma: impact of a dose-modification guideline. Br J Haematol 144:895–903
10. Richardson PG, Sonneveld P, Schuster MW, Irwin D, Stadtmauer EA, Facon T, Harousseau JL, Ben-Yehuda D, Lonial S, Goldschmidt H et al (2005) Bortezomib or high-dose dexamethasone for relapsed multiple myeloma. N Engl J Med 352:2487–2498
11. Richardson PG, Barlogie B, Berenson J, Singhal S, Jagannath S, Irwin D, Rajkumar SV, Hideshima T, Xiao H, Esseltine D et al (2005) Clinical factors predictive of outcome with bortezomib in patients with relapsed, refractory multiple myeloma. Blood 106:2977–2981
12. Richardson PG, Sonneveld P, Schuster MW, Irwin D, Stadtmauer EA, Facon T, Harousseau JL, Ben-Yehuda D, Lonial S, San Miguel JF et al (2007) Safety and efficacy of bortezomib in high-risk and elderly patients with relapsed multiple myeloma. Br J Haematol 137:429–435
13. Jagannath S, Barlogie B, Berenson JR, Singhal S, Alexanian R, Srkalovic G, Orlowski RZ, Richardson PG, Anderson J, Nix D et al (2005) Bortezomib in recurrent and/or refractory multiple myeloma. Initial clinical experience in patients with impared renal function. Cancer 103:1195–1200
14. Chanan-Khan AA, Kaufman JL, Mehta J, Richardson PG, Miller KC, Lonial S, Munshi NC, Schlossman R, Tariman J, Singhal S (2007) Activity and safety of bortezomib in multiple myeloma patients with advanced renal failure: a multicenter retrospective study. Blood 109:2604–2606
15. Ludwig H, Drach J, Graf H, Lang A, Meran JG (2007) Reversal of acute renal failure by bortezomib-based chemotherapy in patients with multiple myeloma. Haematologica 92: 1411–1414
16. San-Miguel JF, Richardson PG, Sonneveld P, Schuster MW, Irwin D, Stadtmauer EA, Facon T, Harousseau JL, Ben-Yehuda D, Lonial S et al (2008) Efficacy and safety of bortezomib in patients with renal impairment: results from the APEX phase 3 study. Leukemia 22:842–849
17. Rajkumar SV, Richardson PG, Hideshima T, Anderson KC (2005) Proteasome inhibition as a novel therapeutic target in human cancer. J Clin Oncol 23:630–639
18. Orlowski RZ, Kuhn DJ (2008) Proteasome inhibitors in cancer therapy: lessons from the first decade. Clin Cancer Res 14:1649–1657
19. Hideshima T, Richardson P, Chauhan D, Palombella VJ, Elliott PJ, Adams J, Anderson KC (2001) The proteasome inhibitor PS-341 inhibits growth, induces apoptosis, and overcomes drug resistance in human multiple myeloma cells. Cancer Res 61:3071–3076
20. Jagannath S, Richardson PG, Barlogie B, Berenson JR, Singhal S, Irwin D, Srkalovic G, Schenkein DP, Esseltine DL, Anderson KC (2006) Bortezomib in combination with dexamethasone for the treatment of patients with relapsed and/or refractory multiple myeloma with less than optimal response to bortezomib alone. Haematologica 91:929–934
21. Mikhael JR, Belch AR, Prince HM, Lucio MN, Maiolino A, Corso A, Petrucci MT, Musto P, Komarnicki M, Stewart AK (2009) High response rate to bortezomib with or without dexamethasone in patients with relapsed or refractory multiple myeloma: results of a global phase 3b expanded access program. Br J Haematol 144:169–175
22. Small GW, Somasundaram S, Moore DT, Shi YY, Orlowski RZ (2003) Repression of mitogen-activated protein kinase (MAPK) phosphatase-1 by anthracyclines contributes to their antiapoptotic activation of p44/42-MAPK. J Pharmacol Exp Ther 307:861–869
23. Small GW, Shi YY, Edmund NA, Somasundaram S, Moore DT, Orlowski RZ (2004) Evidence that mitogen-activated protein kinase phosphatase-1 induction by proteasome inhibitors plays an antiapoptotic role. Mol Pharmacol 66:1478–1490
24. Orlowski RZ, Voorhees PM, Garcia RA, Hall MD, Kudrik FJ, Allred T, Johri AR, Jones PE, Ivanova A, Van Deventer HW et al (2005) Phase 1 trial of the proteasome inhibitor bortezomib and pegylated liposomal doxorubicin in patients with advanced hematologic malignancies. Blood 105:3058–3065

25. Biehn SE, Moore DT, Voorhees PM, Garcia RA, Lehman MJ, Dees EC, Orlowski RZ (2007) Extended follow-up of outcome measures in multiple myeloma patients treated on a phase I study with bortezomib and pegylated liposomal doxorubicin. Ann Hematol 86:211–216
26. Orlowski RZ, Nagler A, Sonneveld P, Blade J, Hajek R, Spencer A, San Miguel J, Robak T, Dmoszynska A, Horvath N et al (2007) Randomized phase III study of pegylated liposomal doxorubicin plus bortezomib compared with bortezomib alone in relapsed or refractory multiple myeloma: combination therapy improves time to progression. J Clin Oncol 25:3892–3901
27. Sonneveld P, Hajek R, Nagler A, Spencer A, Blade J, Robak T, Zhuang SH, Harousseau JL, Orlowski RZ (2008) Combined pegylated liposomal doxorubicin and bortezomib is highly effective in patients with recurrent or refractory multiple myeloma who received prior thalidomide/lenalidomide therapy. Cancer 112:1529–1537
28. Blade J, Sonneveld P, San Miguel JF, Sutherland HJ, Hajek R, Nagler A, Spencer A, Robak T, Cibeira MT, Zhuang SH et al (2008) Pegylated liposomal doxorubicin plus bortezomib in relapsed or refractory multiple myeloma: efficacy and safety in patients with renal function impairment. Clin Lymphoma Myeloma 8:352–355
29. Ma MH, Yang HH, Parker K, Manyak S, Friedman JM, Altamirano C, Wu ZQ, Borad MJ, Frantzen M, Roussos E et al (2003) The proteasome inhibitor PS-341 markedly enhances sensitivity of multiple myeloma tumor cells to chemotherapeutic agents. Clin Cancer Res 9:1136–1144
30. Mitsiades N, Mitsiades CS, Richardson PG, Poulaki V, Tai YT, Chauhan D, Fanourakis G, Gu X, Bailey C, Joseph M et al (2003) The proteasome inhibitor PS-341 potentiates sensitivity of multiple myeloma cells to conventional chemotherapeutic agents: therapeutic applications. Blood 101:2377–2380
31. Richardson P, Jagannath S, Jakubowiak A, Lonial S, Raje N, Alsina M, Ghobrial I, Mazumder A, Munshi N, Vesole DH et al (2008) Lenalidomide, bortezomib, and dexamethasone in patients with relapsed or relapsed/refractory multiple myeloma (MM): encouraging response rates and tolerability with correlation of outcome and adverse cytogenetics in a phase II study. Blood 112: Abstract 1742
32. Mitsiades N, Mitsiades CS, Poulaki V, Chauhan D, Fanourakis G, Gu X, Bailey C, Joseph M, Libermann TA, Treon SP et al (2002) Molecular sequelae of proteasome inhibition in human multiple myeloma cells. Proc Natl Acad Sci USA 99:14374–14379
33. Mitsiades CS, Mitsiades NS, McMullan CJ, Poulaki V, Shringarpure R, Hideshima T, Akiyama M, Chauhan D, Munshi N, Gu X et al (2004) Transcriptional signature of histone deacetylase inhibition in multiple myeloma: biological and clinical implications. Proc Natl Acad Sci USA 101:540–545
34. Hideshima T, Bradner JE, Wong J, Chauhan D, Richardson P, Schreiber SL, Anderson KC (2005) Small-molecule inhibition of proteasome and aggresome function induces synergistic antitumor activity in multiple myeloma. Proc Natl Acad Sci USA 102:8567–8572
35. Nawrocki ST, Carew JS, Pino MS, Highshaw RA, Andtbacka RH, Dunner K Jr, Pal A, Bornmann WG, Chiao PJ, Huang P et al (2006) Aggresome disruption: a novel strategy to enhance bortezomib-induced apoptosis in pancreatic cancer cells. Cancer Res 66:3773–3781
36. Catley L, Weisberg E, Kiziltepe T, Tai YT, Hideshima T, Neri P, Tassone P, Atadja P, Chauhan D, Munshi NC et al (2006) Aggresome induction by proteasome inhibitor bortezomib and alpha-tubulin hyperacetylation by tubulin deacetylase (TDAC) inhibitor LBH589 are synergistic in myeloma cells. Blood 108:3441–3449
37. Badros A, Philip S, Niesvizky R, Goloubeva O, Harris C, Zweibel J, Wright J, Burger A, Grant S, Baer MR et al (2007) Phase I trial of suberoylanilide hydroxamic acid (SAHA) + bortezomib (Bort) in relapsed multiple myeloma (MM) patients (pts). Blood 110: Abstract 1168
38. Weber DM, Jagannath S, Mazumder A, Sobecks R, Schiller GJ, Gavino M, Sumbler C, McFadden C, Chen C, Ricker JL et al (2007) Phase I trial of oral vorinostat (suberoylanilide hydroxamic acid, SAHA) in combination with bortezomib in patients with advanced multiple myeloma. Blood 110: Abstract 1172

39. Weber D, Badros AZ, Jagannath S, Siegel D, Richon V, Rizvi S, Garcia-Vargas J, Reiser D, Anderson KC (2008) Vorinostat plus bortezomib for the treatment of relapsed/refractory multiple myeloma: early clinical experience. Blood 112: Abstract 871
40. Hideshima T, Catley L, Yasui H, Ishitsuka K, Raje N, Mitsiades C, Podar K, Munshi NC, Chauhan D, Richardson PG et al (2006) Perifosine, an oral bioactive novel alkylphospholipid, inhibits Akt and induces in vitro and in vivo cytotoxicity in human multiple myeloma cells. Blood 107:4053–4062
41. Mitsiades CS, Mitsiades NS, McMullan CJ, Poulaki V, Kung AL, Davies FE, Morgan G, Akiyama M, Shringarpure R, Munshi NC et al (2006) Antimyeloma activity of heat shock protein-90 inhibition. Blood 107:1092–1100
42. Voorhees PM, Chen Q, Kuhn DJ, Small GW, Hunsucker SA, Strader JS, Corringham RE, Zaki MH, Nemeth JA, Orlowski RZ (2007) Inhibition of interleukin-6 signaling with CNTO 328 enhances the activity of bortezomib in preclinical models of multiple myeloma. Clin Cancer Res 13:6469–6478
43. Jagannath S, Richardson PG, Sonneveld P, Schuster MW, Irwin D, Stadtmauer EA, Facon T, Harousseau JL, Cowan JM, Anderson KC (2007) Bortezomib appears to overcome the poor prognosis conferred by chromosome 13 deletion in phase 2 and 3 trials. Leukemia 21:151–157
44. Kropff MH, Bisping G, Wenning D, Volpert S, Tchinda J, Berdel WE, Kienast J (2005) Bortezomib in combination with dexamethasone for relapsed multiple myeloma. Leuk Res 29:587–590
45. Sagaster V, Ludwig H, Kaufmann H, Odelga V, Zojer N, Ackermann J, Kuenburg E, Wieser R, Zielinski C, Drach J (2007) Bortezomib in relapsed multiple myeloma: response rates and duration of response are independent of a chromosome 13q-deletion. Leukemia 21:164–168
46. Chang H, Trieu Y, Qi X, Xu W, Stewart KA, Reece D (2007) Bortezomib therapy response is independent of cytogenetic abnormalities in relapsed/refractory multiple myeloma. Leuk Res 31:779–782
47. Hainsworth JD, Spigel DR, Barton J, Farley C, Schreeder M, Hon J, Greco FA (2008) Weekly treatment with bortezomib for patients with recurrent or refractory multiple myeloma: a phase 2 trial of the Minnie Pearl Cancer Research Network. Cancer 113:765–771
48. Suvannasankha A, Smith GG, Juliar BE, Abonour R (2006) Weekly bortezomib/methylprednisolone is effective and well tolerated in relapsed multiple myeloma. Clin Lymphoma Myeloma 7:131–134
49. Reece DE, Rodriguez GP, Chen C, Trudel S, Kukreti V, Mikhael J, Pantoja M, Xu W, Stewart AK (2008) Phase I–II trial of bortezomib plus oral cyclophosphamide and prednisone in relapsed and refractory multiple myeloma. J Clin Oncol 26:4777–4783
50. Chanan-Khan A, Miller KC, Musial L, Padmanabhan S, Yu J, Ailawadhi S, Sher T, Mohr A, Bernstein ZP, Barcos M et al (2009) Bortezomib in combination with pegylated liposomal doxorubicin and thalidomide is an effective steroid independent salvage regimen for patients with relapsed or refractory multiple myeloma: results of a phase II clinical trial. Leuk Lymphoma 50:1096–1101
51. Palumbo A, Ambrosini MT, Benevolo G, Pregno P, Pescosta N, Callea V, Cangialosi C, Caravita T, Morabito F, Musto P et al (2007) Bortezomib, melphalan, prednisone, and thalidomide for relapsed multiple myeloma. Blood 109:2767–2772
52. Wolf J, Richardson PG, Schuster M, LeBlanc A, Walters IB, Battleman DS (2008) Utility of bortezomib retreatment in relapsed or refractory multiple myeloma patients: a multicenter case series. Clin Adv Hematol Oncol 6:755–760
53. Conner TM, Doan QD, Walters IB, LeBlanc AL, Beveridge RA (2008) An observational, retrospective analysis of retreatment with bortezomib for multiple myeloma. Clin Lymphoma Myeloma 8:140–145
54. Hrusovsky I, Emmerich B, von Rohr A, Engelhardt M, Voegeli J, Taverna C, Olie R, Pliskat H, Frohn C, Hess G (2008) Bortezomib retreatment in relapsed multiple myeloma (MM): results from a binational, multicenter retrospective survey. Blood 112: Abstract 2775

55. Rubio-Martinez A, Recasens V, Soria B, Montañes MA, Rubio-Escuin R, Giraldo P (2008) Response to re-treatment on relapse multiple myeloma patients previously treated with bortezomib. Haematologica 93: Abstract 649
56. Ciolli S, Leoni F, Casini C, Breschi C, Santini V, Bosi A (2008) The addition of liposomal doxorubicin to bortezomib, thalidomide and dexamethasone significantly improves clinical outcome of advanced multiple myeloma. Br J Haematol 141:814–819
57. Palumbo A, Gay F, Bringhen S, Falcone A, Pescosta N, Callea V, Caravita T, Morabito F, Magarotto V, Ruggeri M et al (2008) Bortezomib, doxorubicin and dexamethasone in advanced multiple myeloma. Ann Oncol 19:1160–1165
58. Fenk R, Michael M, Zohren F, Graef T, Czibere A, Bruns I, Neumann F, Fenk B, Haas R, Kobbe G (2007) Escalation therapy with bortezomib, dexamethasone and bendamustine for patients with relapsed or refractory multiple myeloma. Leuk Lymphoma 48:2345–2351
59. Berenson JR, Yang HH, Sadler K, Jarutirasarn SG, Vescio RA, Mapes R, Purner M, Lee SP, Wilson J, Morrison B et al (2006) Phase I/II trial assessing bortezomib and melphalan combination therapy for the treatment of patients with relapsed or refractory multiple myeloma. J Clin Oncol 24:937–944
60. Berenson JR, Yang HH, Vescio RA, Nassir Y, Mapes R, Lee SP, Wilson J, Yellin O, Morrison B, Hilger J et al (2008) Safety and efficacy of bortezomib and melphalan combination in patients with relapsed or refractory multiple myeloma: updated results of a phase 1/2 study after longer follow-up. Ann Hematol 87:623–631
61. Popat R, Oakervee H, Williams C, Cook M, Craddock C, Basu S, Singer C, Harding S, Foot N, Hallam S et al (2009) Bortezomib, low-dose intravenous melphalan, and dexamethasone for patients with relapsed multiple myeloma. Br J Haematol 144:887–894
62. Terpos E, Kastritis E, Roussou M, Heath D, Christoulas D, Anagnostopoulos N, Eleftherakis-Papaiakovou E, Tsionos K, Croucher P, Dimopoulos MA (2008) The combination of bortezomib, melphalan, dexamethasone and intermittent thalidomide is an effective regimen for relapsed/refractory myeloma and is associated with improvement of abnormal bone metabolism and angiogenesis. Leukemia 22:2247–2256
63. Davies FE, Wu P, Jenner M, Srikanth M, Saso R, Morgan GJ (2007) The combination of cyclophosphamide, velcade and dexamethasone induces high response rates with comparable toxicity to velcade alone and velcade plus dexamethasone. Haematologica 92:1149–1150
64. Kropff M, Bisping G, Schuck E, Liebisch P, Lang N, Hentrich M, Dechow T, Kroger N, Salwender H, Metzner B et al (2007) Bortezomib in combination with intermediate-dose dexamethasone and continuous low-dose oral cyclophosphamide for relapsed multiple myeloma. Br J Haematol 138:330–337
65. Pineda-Roman M, Zangari M, van Rhee F, Anaissie E, Szymonifka J, Hoering A, Petty N, Crowley J, Shaughnessy J, Epstein J et al (2008) VTD combination therapy with bortezomib-thalidomide-dexamethasone is highly effective in advanced and refractory multiple myeloma. Leukemia 22:1419–1427
66. Berenson JR, Yellin O, Patel R, Duvivier H, Nassir Y, Mapes R, Abaya CD, Swift RA (2009) A phase I study of samarium lexidronam/bortezomib combination therapy for the treatment of relapsed or refractory multiple myeloma. Clin Cancer Res 15:1069–1075
67. Richardson P, Wolf J, Jakubowiak A, Zonder J, Lonial S, Irwin DH, Densmore J, Krishnan A, Raje N, Bar MH et al (2008) Phase I/II results of a multicenter trial of perifosine (KRX-0401) + bortezomib in patients with relapsed or relapsed/refractory multiple myeloma who were previously relapsed from or refractory to bortezomib. Blood 112: Abstract 870
68. Richardson P, Chanan-Khan AA, Lonial S, Krishnan A, Alsina M, Carroll M, Adler K, Cropp7, G, Mitsiades C, Johnson R et al (2006) A multicenter phase 1 clinical trial of tanespimycin (KOS-953) + bortezomib (BZ): encouraging activity and manageable toxicity in heavily pre-treated patients with relapsed refractory multiple myeloma (MM). Blood 108: Abstract 406
69. Siegel DD, Sezer O, San Miguel JF, Mateos M-V, Prosser I, Cavo M, Jalaluddin M, Hazell K, Bourquelot PM, Anderson KC (2008) A phase IB, multicenter, open-label, dose-escalation

study of oral panobinostat (LBH589) and i.v. bortezomib in patients with relapsed multiple myeloma. Blood 112: Abstract 2781
70. Rossi J-F, Manges RF, Sutherland HJ, Jagannath S, Voorhees P, Sonneveld P, Delforge M, Pegourie B, Alegre A, de la Rubia J et al (2008) Preliminary results of CNTO 328, an anti-interleukin-6 monoclonal antibody, in combination with bortezomib in the treatment of relapsed or refractory multiple myeloma. Blood 112: Abstract 867
71. Berenson JR, Matous J, Swift RA, Mapes R, Morrison B, Yeh HS (2007) A phase I/II study of arsenic trioxide/bortezomib/ascorbic acid combination therapy for the treatment of relapsed or refractory multiple myeloma. Clin Cancer Res 13:1762–1768
72. Ghobrial IM, Munshi N, Schlossman R, Chuma S, Leduc R, Nelson M, Sam A, O'Connor K, Harris B, Warren D et al (2008) Phase I trial of CCI-779 (temsirolimus) and weekly bortezomib in relapsed and/or refractory multiple myeloma. Blood 112: Abstract 3696

Bortezomib-Induced Peripheral Neuropathy in Multiple Myeloma: Principles of Identification and Management

Jacob P. Laubach and Paul G. Richardson

Abstract The first-in-class proteasome inhibitor bortezomib is a potent anti-myeloma drug that now figures prominently in the management of patients with both newly diagnosed and relapsed multiple myeloma. With current studies evaluating its efficacy as part of conditioning prior to and maintenance therapy following autologous stem cell transplantation, it is likely that its role in the management of the disease will expand further. Peripheral neuropathy was recognized as a toxicity associated with bortezomib at an early point in the clinical development of the agent, and further study has provided considerable insight regarding this issue. It is critical that clinicians recognize and appropriately manage bortezomib-induced PN to optimize care for patients. This chapter focuses on the identification, characterization, and management of bortezomib-induced peripheral neuropathy.

1 Introduction

As comprehensively reviewed in other chapters, bortezomib is a key part of the therapeutic armamentarium utilized in the treatment of multiple myeloma (MM). Its efficacy in the context of relapsed and refractory as well as newly diagnosed MM has been clearly demonstrated in large, randomized clinical trials [1, 2]. The agent is currently being evaluated in other settings relevant to the care of MM patients as well, including conditioning prior to autologous stem cell transplantation (ASCT) and maintenance therapy following ASCT [3, 4]. As use of bortezomib in MM expands, patients will receive higher cumulative doses of the drug and thus will be potentially more susceptible to associated toxicities. Bortezomib-induced peripheral neuropathy (PN) is of particular concern in this respect, as it is known to be a dose-dependent phenomenon although not necessarily cumulative. This chapter

J.P. Laubach (✉) and P.G. Richardson
Dana-Farber Cancer Institute, Boston, MA 02115, USA
e-mail: JacobP_Laubach@dfci.harvard.edu, Paul_Richardson@dfci.harvard.edu

I.M. Ghobrial et al. (eds.), *Bortezomib in the Treatment of Multiple Myeloma*, Milestones in Drug Therapy,
DOI 10.1007/978-3-7643-8948-2_7,

focuses on the identification, characterization, and management of bortezomib-induced PN (also known as Bipn), which is a critical aspect of optimizing care for MM patients who receive this important therapeutic agent.

2 Clinical Trial Experience: Observations and Lessons

Treatment-associated PN occurred in 5 of 27 patients (19%) enrolled in the initial phase I study of single-agent bortezomib for advanced hematologic malignancies by Orlowski and colleagues [5]. PN was felt to be related to bortezomib in three of five patients, and was dose-limiting (grade 3/4) in one patient. Confirmatory evidence of a causal relationship between bortezomib and PN came from phase II and III studies of single-agent bortezomib in relapsed MM, wherein the overall rate of bortezomib-associated PN ranged from 31 to 41%, and the rate of high grade PN (grade 3) from 8 to 15% [1, 6, 7].

The incidence of bortezomib-associated PN has been similar in studies involving patients with newly diagnosed disease. In a phase III study comparing melphalan and prednisone (MP) with bortezomib plus MP (VMP) among patients with previously untreated MM, the overall rate of PN was 44% in the VMP arm (vs. 5% with MP) and rate of grade 3/4 PN 14% (vs. none with MP) [2]. Meanwhile, in a retrospective comparison of patients who received bortezomib as part of therapy for newly diagnosed vs. relapsed MM, there was no difference in the incidence, severity, or outcome (as reflected by improvement or resolution of symptoms) of treatment-associated PN [8]. However, patients who received bortezomib-containing therapy as initial induction did experience less neuropathic pain and symptoms, which resolved and/or improved more quickly than in those with relapsed disease.

Interestingly, results of early phase clinical trials suggest that the incidence of severe therapy-associated PN may be lower when bortezomib is used in combination with the immunomodulatory drugs thalidomide or lenalidomide (IMiDs). In a phase I/II study of bortezomib plus thalidomide and dexamethasone (VTD) in relapsed and refractory MM, there were no instances of grade 3 or higher PN at a bortezomib dose of 1.0 mg/m^2 [9]. At the higher bortezomib dose of 1.3 mg/m^2, three patients (8%) developed grade 3 PN. In a phase I study of bortezomib in conjunction with lenalidomide and dexamethasone (RVD) in relapsed and refractory MM, the rate of therapy-related PN was 42%, but there were no grade 3/4 events [10].

3 Clinical Characteristics of Bortezomib-Induced PN

Bortezomib-induced PN typically produces sensory symptoms such as numbness, tingling, or diminished sensation of external stimuli. Dyesthesia, the abnormal perception of touch or other stimuli, can also occur. Pain develops in some patients as well, and manifests variably as a burning, lancinating, or cold sensation. The lower extremities and especially the feet are often affected first, with upper extremity involvement occurring later. PN symptoms are usually symmetric, and the presence of asymmetric symptoms should prompt consideration of a possible coexisting condition such as carpal tunnel syndrome or underlying radiculopathy. In more advanced cases, motor and/or autonomic function may be affected although motor neuropathy is rare. Autonomic nerve dysfunction frequently manifests as gastrointestinal symptoms such as abdominal distension and pain related to gastric dysmotility, and less commonly as orthostatic hypotension.

Many patients with MM have preexisting PN due to both underlying disease-related nerve damage and the effects of prior neurotoxic chemotherapeutic agents, thus being potentially more susceptible to the neuropathic effects of bortezomib [11–13]. However, it is emphasized that the severity of treatment-emergent PN does not necessarily correlate with the presence of preexisting PN, while conversely, patients without preexisting PN can develop symptoms [14]. Older age, preexisting peripheral nervous system dysfunction, prior thalidomide exposure, extensive bortezomib exposure, and underlying diabetes mellitus appear to be predictors for the development of bortezomib-associated PN in some studies, but not others [8, 13, 15, 16].

Although a small proportion of patients develop PN early in the course of bortezomib therapy, most do so after three or more cycles of treatment, with few experiencing PN after cycles five or six, consistent with the observation that the incidence of bortezomib-induced PN increases with higher doses of bortezomib but appears to plateau at a dose of approximately 30 mg/m^2 and is not cumulative [11, 17]. Formal nerve conduction studies (NCS) may reveal a decline in amplitude of sensory nerve action potentials, while motor action potentials tend not to be affected [11]. NCS most often suggests axonal injury, although evidence of demyelination has been reported in a subset of patients as well, and this may be a nonspecific late effect vs. a feature of bortezomib induced injury per se [18, 19].

In contrast to the PN observed in association with thalidomide, bortezomib-induced PN is most often reversible. Improvement and/or resolution of PN typically occur within the first 2 or 3 months following treatment discontinuation, but in some cases occurs more slowly and only resolves gradually over the course of 12 months or more [11, 12, 15].

4 Potential Mechanisms of Bortezomib-Induced Peripheral Neuropathy

Within the peripheral nervous system, sensory nerves are small in diameter and either thinly myelinated or unmyelinated, whereas motor neurons are larger and myelinated [20]. The smaller, unmyelinated sensory fibers appear to be more prone to bortezomib-induced injury. Moreover, the cell bodies of sensory peripheral nerves reside in the dorsal root ganglion (DRG), and appear to be particularly susceptible to the neurotoxic effects of bortezomib as well as other potential neurotoxins due to the absence of the blood–brain barrier in this location [21].

Rat models have been utilized to investigate mechanisms of bortezomib-induced nerve injury and provide important insights related to this phenomenon. In one study, bortezomib administration resulted in a reversible decrease in sensory nerve conduction velocity, which correlated with specific pathologic changes in tissue specimens, including mitochondrial and endoplasmic reticulum damage within the DRG, injury to myelin-producing Schwann cells, and axonal degeneration [22]. In another study, axonal damage involving unmyelinated nerve fibers was observed in response to bortezomib administration, while DRG neurons and satellite cells remained intact [23]. A third study characterized physiologic changes occurring within the nuclei of DRG neurons following bortezomib administration, including the aggregation of ubiquitin-protein conjugates, reduction of transcriptional activity, and retention of poly(A) RNA and its bindings proteins poly(A) binding protein nuclear 1 and Sam 68 in dense granules [24]. This suggested that bortezomib inhibits transcription and translation in DRG neurons. Together, the findings of these studies suggest that various mechanisms within the peripheral nervous system are implicated in bortezomib-induced nerve injury, primarily affects the DRG, and the axon itself, with other downstream effects (such as on the myelin steaty also involved.

In vitro studies utilizing MM and other cell lines have also provided insights regarding possible mechanisms. Specifically, dysregulation of mitochondrial calcium homeostasis has been proposed as a mediator of bortezomib-associated neurotoxicity [25–27]. Bortezomib has also been shown in cultured cells to increase tubulin polymerization and stabilization in association with a significant increase in microtubule associated proteins [28]. It is likely that the cellular mechanisms underlying bortezomib-induced neurotoxicity will be further elucidated with ongoing analyses, both preclinically and clinically.

5 Evaluation and Monitoring

The importance of ongoing clinical evaluation by both history and physical examination throughout the course of bortezomib therapy is vital. Attention should be given at the onset of therapy to comorbidities such as diabetes, prior alcohol use, nutritional deficiencies and other conditions potentially associated with underlying PN. As therapy continues, the patient should be queried regarding symptoms commonly associated with bortezomib-induced PN such as parasthesias and dysesthesia, as well as gastrointestinal symptoms such as bloating or nausea and less common but important manifestations such as postural dizziness/light-headedness. The examination of patients receiving bortezomib should include regular evaluation of sensory function, deep tendon reflexes, motor capacity, and balance.

Symptoms are graded on a scale of 0 (no symptoms) to 4 (severe symptoms) using a standardized grading system. Several grading systems exist, including the World Health Organization (WHO), Eastern Cooperative Group (ECOG), National Cancer Institute Common Terminology Criteria for Adverse Events [CTC]), and Ajani Sensory Systems [29–32]. The Total Neuropathy Score has recently been developed in an effort to improve the sensitivity for detecting changes in the degree of PN [33, 34]. Interobserver variability has been shown between the grading symptoms, such that consistent use of a single grading system throughout a given patient's management is recommended [35], and especially in the context of clinical trials.

Patient self-assessment forms are also available and may prove helpful in assessing the nature and extent of symptoms related to bortezomib-induced PN. Given the subjective nature of symptoms associated with drug-induced PN, self-assessment tools may increase the accuracy and consistency of patient reporting. This is important in the management of bortezomib-induced PN, since interventions aimed at preventing progression of symptoms are based on the presence and severity of symptoms. Available patient-based evaluation systems include the Functional Assessment of Cancer Therapy (FACT)/Gynecology Oncology Group-Neurotoxicity [36] and Patient Neurotoxicity Questionnaire (PNQ) [37].

Formal neurophysiological testing can also be undertaken when necessary as a supplement to clinical evaluation. Electromyography (EMG) and/or NCS provide objective measurements of nerve injury, which are particularly valuable in assessing patients with atypical PN with respect to the quality, severity, or time course of symptoms. It is noted, however, that a discrepancy between patient-reported symptoms and formal neurophysiologic testing can occur, and that treatment decisions are best based on findings from the clinical assessment including patient-reported symptom [12, 38, 39].

6 Management

If bortezomib-induced PN develops during the course of therapy, interventions are taken to both prevent further progression of symptoms and manage existing symptoms. On the basis of the observations of an association between bortezomib and PN in the initial clinical development of the agent, a dose reduction schema program was implemented in the phase III Apex trial [1, 6, 7, 11]. As shown in Table 1, dose reduction from 1.3 to 1.0 mg/m^2 is recommended for grade 1 symptoms with pain or grade 2 PN. Bortezomib therapy is held for grade 2 with pain or grade 3 symptoms until resolution of toxicity, with reinitiation of therapy at a reduced dose on a weekly rather than bi-weekly schedule. Therapy is discontinued for grade 4 PN. Retrospective comparison of the APEX trial and earlier trials suggest that a dose reduction guideline based on the presence of PN resulted in a significantly lower rate of high grade symptoms [39]. Appropriate dose modification or discontinuation of therapy as necessary based on symptoms thus represents a critical element in the management of MM patients receiving bortezomib-containing therapies.

Table 1 Dose modification guideline for bortezomib-related neuropathic pain and/or peripheral sensory or motor neuropathy

Severity of peripheral neuropathy signs and symptoms	Modification of bortezomib dose and regimen
Grade 1 (paresthesias, weakness and/or loss of reflexes) without pain or loss of function	No action
Grade 1 with pain or Grade 2 (interfering with function but not with activities of daily living)	Reduce to 1.0 mg/m^2
Grade 2 with pain or Grade 3 (interfering with activities of daily living)	Withhold treatment until toxicity resolves, then reinitiate at a dose of 0.7 mg/m^2 once weekly
Grade 4 (sensory neuropathy that is disabling or motor neuropathy that is life-threatening or leads to paralysis)	Discontinue

Grading for this currently recommended dose modification guideline is based on National Cancer Institute Common Terminology Criteria for Adverse Events (NCI CTCAE) version 3.0. In APEX, the dose modification guideline used was the same, but based on NCI CTC version 2.0 grading; in addition, patients experiencing grade 3 peripheral neuropathy with pain were to discontinue bortezomib

Despite the preventive strategies aimed at reducing the frequency and severity of bortezomib-induced PN, a significant number of individuals receiving the agent ultimately require pharmacologic interventions for symptom control. Effective options in this respect include gabapentin and pregabalin, both of which are believed to inhibit the release of excitatory neurotransmitters involved in nociception; the tricyclic antidepressants amitriptyline and desipramine; the serotonin and norepinephrine reuptake inhibitor duloxetine; and opioid agents such as oxycodone, morphine sulfate, methadone, and hydrocodone [40]. These agents are initiated at a low dose when necessary, with an increase in dose based on the degree of symptom relief. It is also noted that the application of topical agents such as menthol, which activates the transient receptor potential melastatin (TRPM8) receptor, as well as other emollients such as cocoa butter have been shown in some cases to provide significant symptom relief in the setting of bortezomib-induced PN [41, 12]. Nutritional supplevents may have a role for randomized controlled swchies are lacking; current guitelines suggting caution with concomitance an dargs of bortezomib administration [11, 12] allow with use at other times to allter symptoms has been reported to be helpful

7 Future Directions

Because bortezomib plays a critical role in the overall management of MM, the development of strategies to both prevent and manage therapy-associated PN is a matter of great importance. Modifications in the dose and schedule of bortezomib administration have recently been evaluated in several clinical trials; encouraging preliminary results from these studies have been reported showing substantial reductions in PN and no loss of clinical benefit and data from long-term follow up regarding both toxicity and efficacy are thus awaited with great interest. Specifically in one study by Mateos and colleagues, bortezomib was administered as part of either VMP or VTD, using the conventional biweekly schedule for two 3-week cycles, then using a weekly schedule for five 5-week cycles, followed by maintenance chemotherapy with bortezomib twice weekly in a 3-week cycle every 3 months [42]. The concept employed in this trial is similar to the strategy of maintenance bortezomib therapy utilized in the phase III VISTA trial, wherein patients received weekly bortezomib beginning with cycle five and extending through completion of protocol directed therapy [2]. However, dose reduction in the study by Mateos et al. occurred at an earlier point in treatment with a reduced rate of PN subsequently reported. In a phase II study by Reeder and colleagues, patients with untreated MM received bortezomib in conjunction with cyclophosphamide and dexamethasone (CYBORD), with bortezomib given at the standard dose (1.3 mg/m^2) on days 1, 4, 8, and 11 of a 28 day cycle or weekly at 1.5 mg/m^2, again with less PN seen with the weekly schedule compared to the twice weekly regines [43]. The results of studies such as these will determine whether modifications in dose and schedule of bortezomib administration decrease the incidence of treatment-associated PN while maintaining therapeutic efficacy, and initial results certainly seem to suggest this, especially when bortezomib is used in combination.

It may be that adjunctive therapies with neuroprotective characteristics during the course of bortezomib-containing therapy will alter the natural history of treatment-associated PN. Although such agents have not yet been evaluated in MM patients receiving bortezomib, they have been evaluated in PN occurring in other clinical settings. For example, the low molecular weight metabolic antioxidant α-lipoic acid has been evaluated and shown to have benefit in diabetic neuropathy [44–46]. The amino acid glutamine, which plays an important role in purine and pyrimidine synthesis, has been shown to have a neuroprotective effect in breast cancer patients receiving paclitaxel [47] and in patients with colorectal cancer receiving oxaliplatin [48]. Similarly, the amino acid L-carnitine, which traffics free fatty acids into the mitochondria, has been shown to reduce PN-associated symptoms among diabetic patients and those receiving paclitaxel chemotherapy [49]. It appears that vitamin B supplementation may offer modest benefit in relieving PN associated with either alcoholism or diabetes mellitus [50]. In addition, there is suggestion from a randomized phase II trial that vitamin E supplementation during paclitaxel-based chemotherapy reduces the rate of chemotherapy-induced PN [51]. It is important for nutritional supplements to be evaluated prior to their standard use in MM patients, as such agents may have unanticipated effects. For example, high doses of vitamin B6 (pyridoxine) have been associated with the development of sensorimotor neuropathy [52]. Moreover, several recent in vitro studies have demonstrated an inhibitory effect of vitamin C on bortezomib [53, 54]. Anecdotal experience and some prospective but uncontrolled studies suggest that benefit from these supplements does occur, but as mentioned previously randomized trials are needed [11, 12].

Second generation proteasome inhibitors such as carlfilzomib and NPI0052 may exhibit distinct toxicity profiles, with a lower incidence of PN [55, 56]. Development of the second generation IMiDs such as lenalidomide and pomalidomide, which in comparison to thalidomide are associated with less PN, provides a precedent in this respect that may ultimately also apply to the arena of proteasome inhibition [57, 58]. Finally, certain combinations of chemotherapeutic agents such as tanespimycin and bortezomib, which upregulation HSP70 as part of HSP90 inhibition, may abrogate neurotoxicity and so enhance the neuropeutic index of such combination approaches [59, 60].

References

1. Richardson PG, Sonneveld P, Schuster MW et al (2005) Bortezomib or high-dose dexamethasone for relapsed multiple myeloma. N Engl J Med 352:2487–2498
2. San Miguel JF, Schlag R, Khuageva NK et al (2008) Bortezomib plus melphalan and prednisone for initial treatment of multiple myeloma. N Engl J Med 359:906–917
3. Roussel M, Moreau P, Huynh A et al (2010) Bortezomib and high-dose melphalan as conditioning regimen before autologous stem cell transplantation in patients with de novo multiple myeloma: a phase 2 study of the Intergroupe Francophone du Myelome (IFM). Blood 115:32–37
4. Magarotto V, Palumbo A (2009) Evolving role of novel agents for maintenance therapy in myeloma. Cancer J 15:494–501
5. Orlowski RZ, Stinchcombe TE, Mitchell BS et al (2002) Phase I trial of the proteasome inhibitor PS-341 in patients with refractory hematologic malignancies. J Clin Oncol 20:4420–4427
6. Richardson PG, Barlogie B, Berenson J et al (2003) A phase 2 study of bortezomib in relapsed, refractory myeloma. N Engl J Med 348:2609–2617
7. Jagannath S, Barlogie B, Berenson J et al (2004) A phase 2 study of two doses of bortezomib in relapsed or refractory myeloma. Br J Haematol 127:165–172
8. Corso A, Mangiacavalli S, Varettoni M, Pascutto C, Zappasodi P, Lazzarino M (2010) Bortezomib-induced peripheral neuropathy in multiple myeloma: a comparison between previously treated and untreated patients. Leuk Res 34(4):471–474
9. Pineda-Roman M, Zangari M, van Rhee F et al (2008) VTD combination therapy with bortezomib-thalidomide-dexamethasone is highly effective in advanced and refractory multiple myeloma. Leukemia 22:1419–1427
10. Richardson PG, Weller E, Jagannath S et al (2009) Multicenter, phase I, dose-escalation trial of lenalidomide plus bortezomib for relapsed and relapsed/refractory multiple myeloma. J Clin Oncol 27(34):5713–5719
11. Richardson PG, Briemberg H, Jagannath S et al (2006) Frequency, characteristics, and reversibility of peripheral neuropathy during treatment of advanced multiple myeloma with bortezomib. J Clin Oncol 24:3113–3120
12. Richardson PG, Xie W, Mitsiades C et al (2009) Single-agent bortezomib in previously untreated multiple myeloma: efficacy, characterization of peripheral neuropathy, and molecular correlations with response and neuropathy. J Clin Oncol 27:3518–3525
13. Stubblefield MD, Slovin S, MacGregor-Cortelli B et al (2006) An electrodiagnostic evaluation of the effect of pre-existing peripheral nervous system disorders in patients treated with the novel proteasome inhibitor bortezomib. Clin Oncol (R Coll Radiol) 18:410–418
14. Lanzani F, Mattavelli L, Frigeni B et al (2008) Role of a pre-existing neuropathy on the course of bortezomib-induced peripheral neurotoxicity. J Peripher Nerv Syst 13:267–274

15. El-Cheikh J, Stoppa AM, Bouabdallah R et al (2008) Features and risk factors of peripheral neuropathy during treatment with bortezomib for advanced multiple myeloma. Clin Lymphoma Myeloma 8:146–152
16. Badros A, Goloubeva O, Dalal JS et al (2007) Neurotoxicity of bortezomib therapy in multiple myeloma: a single-center experience and review of the literature. Cancer 110:1042–1049
17. Berenson JR, Jagannath S, Barlogie B et al (2005) Safety of prolonged therapy with bortezomib in relapsed or refractory multiple myeloma. Cancer 104:2141–2148
18. Chaudhry V, Cornblath DR, Polydefkis M, Ferguson A, Borrello I (2008) Characteristics of bortezomib- and thalidomide-induced peripheral neuropathy. J Peripher Nerv Syst 13:275–282
19. Filosto M, Rossi G, Pelizzari AM et al (2007) A high-dose bortezomib neuropathy with sensory ataxia and myelin involvement. J Neurol Sci 263:40–43
20. Stubblefield MD, Burstein HJ, Burton AW et al (2009) NCCN task force report: management of neuropathy in cancer. J Natl Compr Canc Netw 7(Suppl 5):S1–S26, quiz S27–S28
21. Bigotte L, Arvidson B, Olsson Y (1982) Cytofluorescence localization of adriamycin in the nervous system. I. Distribution of the drug in the central nervous system of normal adult mice after intravenous injection. Acta Neuropathol 57:121–129
22. Cavaletti G, Gilardini A, Canta A et al (2007) Bortezomib-induced peripheral neurotoxicity: a neurophysiological and pathological study in the rat. Exp Neurol 204:317–325
23. Meregalli C, Canta A, Carozzi VA et al (2010) Bortezomib-induced painful neuropathy in rats: a behavioral, neurophysiological and pathological study in rats. Eur J Pain 14(4):343–350
24. Casafont I, Berciano MT, Lafarga M (2010) Bortezomib induces the formation of nuclear poly(A) RNA granules enriched in Sam68 and PABPN1 in sensory ganglia neurons. Neurotox Res 17:167–178
25. Pei XY, Dai Y, Grant S (2003) The proteasome inhibitor bortezomib promotes mitochondrial injury and apoptosis induced by the small molecule Bcl-2 inhibitor HA14-1 in multiple myeloma cells. Leukemia 17:2036–2045
26. Pei XY, Dai Y, Grant S (2004) Synergistic induction of oxidative injury and apoptosis in human multiple myeloma cells by the proteasome inhibitor bortezomib and histone deacetylase inhibitors. Clin Cancer Res 10:3839–3852
27. Landowski TH, Megli CJ, Nullmeyer KD, Lynch RM, Dorr RT (2005) Mitochondrial-mediated disregulation of Ca2+ is a critical determinant of Velcade (PS-341/bortezomib) cytotoxicity in myeloma cell lines. Cancer Res 65:3828–3836
28. Poruchynsky MS, Sackett DL, Robey RW, Ward Y, Annunziata C, Fojo T (2008) Proteasome inhibitors increase tubulin polymerization and stabilization in tissue culture cells: a possible mechanism contributing to peripheral neuropathy and cellular toxicity following proteasome inhibition. Cell Cycle 7:940–949
29. Miller AB, Hoogstraten B, Staquet M, Winkler A (1981) Reporting results of cancer treatment. Cancer 47:207–214
30. Ajani JA, Welch SR, Raber MN, Fields WS, Krakoff IH (1990) Comprehensive criteria for assessing therapy-induced toxicity. Cancer Invest 8:147–159
31. Oken MM, Creech RH, Tormey DC et al (1982) Toxicity and response criteria of the Eastern Cooperative Oncology Group. Am J Clin Oncol 5:649–655
32. Institute NC (2006) National Cancer Institute-Common Terminology Criteria for Adverse Events (NCI-CTCAE), version 3.0. Available at: http://ctep.cancer.gov/protocolDevelopment/electronic_applications/docs/ctcaev3.pdf.
33. Cavaletti G, Jann S, Pace A et al (2006) Multi-center assessment of the total neuropathy score for chemotherapy-induced peripheral neurotoxicity. J Peripher Nerv Syst 11:135–141
34. Cavaletti G, Frigeni B, Lanzani F et al (2007) The Total Neuropathy Score as an assessment tool for grading the course of chemotherapy-induced peripheral neurotoxicity: comparison with the National Cancer Institute-Common Toxicity Scale. J Peripher Nerv Syst 12:210–215

35. Postma TJ, Heimans JJ, Muller MJ, Ossenkoppele GJ, Vermorken JB, Aaronson NK (1998) Pitfalls in grading severity of chemotherapy-induced peripheral neuropathy. Ann Oncol 9:739–744
36. Calhoun EA, Welshman EE, Chang CH et al (2003) Psychometric evaluation of the Functional Assessment of Cancer Therapy/Gynecologic Oncology Group-Neurotoxicity (Fact/GOG-Ntx) questionnaire for patients receiving systemic chemotherapy. Int J Gynecol Cancer 13:741–748
37. Hausheer FH, Schilsky RL, Bain S, Berghorn EJ, Lieberman F (2006) Diagnosis, management, and evaluation of chemotherapy-induced peripheral neuropathy. Semin Oncol 33:15–49
38. du Bois A, Schlaich M, Luck HJ et al (1999) Evaluation of neurotoxicity induced by paclitaxel second-line chemotherapy. Support Care Cancer 7:354–361
39. Richardson PG, Sonneveld P, Schuster MW et al (2009) Reversibility of symptomatic peripheral neuropathy with bortezomib in the phase III APEX trial in relapsed multiple myeloma: impact of a dose-modification guideline. Br J Haematol 144:895–903
40. Cleary JF (2007) The pharmacologic management of cancer pain. J Palliat Med 10:1369–1394
41. Colvin LA, Johnson PR, Mitchell R, Fleetwood-Walker SM, Fallon M (2008) From bench to bedside: a case of rapid reversal of bortezomib-induced neuropathic pain by the TRPM8 activator, menthol. J Clin Oncol 26:4519–4520
42. Mateos M, Oriol A, Martinez J et al (2009) A prospective, multicenter, randomized trial of bortezomib/melphalan/prednisone (VMP) versus bortezomib/thalidomide/prednisone (VTP) as induction therapy followed by maintenance treatment with bortezomib/thalidomide (VT) versus bortezomib/prednisone (VP) in elderly untreated patients with multiple myeloma older than 65 years. Abstract 3. Blood 114:3
43. Reeder C, Reece D, Kukreti V et al (2009) A phase II trial comparison of once versus twice weekly bortezomib in CYBORD chemotherapy for newly diagnosed myeloma: identical high response rates and less toxicity. Abstract 616. Blood 114:255
44. Ziegler D, Hanefeld M, Ruhnau KJ et al (1995) Treatment of symptomatic diabetic peripheral neuropathy with the anti-oxidant alpha-lipoic acid. A 3-week multicentre randomized controlled trial (ALADIN Study). Diabetologia 38:1425–1433
45. Packer L, Kraemer K, Rimbach G (2001) Molecular aspects of lipoic acid in the prevention of diabetes complications. Nutrition 17:888–895
46. Ziegler D, Hanefeld M, Ruhnau KJ et al (1999) Treatment of symptomatic diabetic polyneuropathy with the antioxidant alpha-lipoic acid: a 7-month multicenter randomized controlled trial (ALADIN III Study). ALADIN III Study Group. Alpha-Lipoic Acid in Diabetic Neuropathy. Diabetes Care 22:1296–1301
47. Vahdat L, Papadopoulos K, Lange D et al (2001) Reduction of paclitaxel-induced peripheral neuropathy with glutamine. Clin Cancer Res 7:1192–1197
48. Wang WS, Lin JK, Lin TC et al (2007) Oral glutamine is effective for preventing oxaliplatin-induced neuropathy in colorectal cancer patients. Oncologist 12:312–319
49. Jin HW, Flatters SJ, Xiao WH, Mulhern HL, Bennett GJ (2008) Prevention of paclitaxel-evoked painful peripheral neuropathy by acetyl-L-carnitine: effects on axonal mitochondria, sensory nerve fiber terminal arbors, and cutaneous Langerhans cells. Exp Neurol 210:229–237
50. Ang CD, Alviar MJ, Dans AL et al (2008) Vitamin B for treating peripheral neuropathy. Cochrane Database Syst Rev Issue 3:CD004573
51. Argyriou AA, Chroni E, Koutras A et al (2006) Preventing paclitaxel-induced peripheral neuropathy: a phase II trial of vitamin E supplementation. J Pain Symptom Manage 32:237–244
52. Gdynia HJ, Muller T, Sperfeld AD et al (2008) Severe sensorimotor neuropathy after intake of highest dosages of vitamin B6. Neuromuscul Disord 18:156–158
53. Zou W, Yue P, Lin N et al (2006) Vitamin C inactivates the proteasome inhibitor PS-341 in human cancer cells. Clin Cancer Res 12:273–280
54. Perrone G, Hideshima T, Ikeda H et al (2009) Ascorbic acid inhibits antitumor activity of bortezomib in vivo. Leukemia 23:1679–1686

55. O'Connor OA, Stewart AK, Vallone M et al (2009) A phase 1 dose escalation study of the safety and pharmacokinetics of the novel proteasome inhibitor carfilzomib (PR-171) in patients with hematologic malignancies. Clin Cancer Res 15:7085–7091
56. Richardson P, Hofmeister C, Jakubowiak A et al (2009) Phase 1 clinical trial of the novel structure proteasome inhibitor NPI-0052 in patients with relapsed and relapsed/refractory multiple myeloma (MM). Abstract 431. Blood 114:179
57. Lacy MQ, Hayman SR, Gertz MA et al (2009) Pomalidomide (CC4047) plus low-dose dexamethasone as therapy for relapsed multiple myeloma. J Clin Oncol 27:5008–5014
58. Dimopoulos M, Spencer A, Attal M et al (2007) Lenalidomide plus dexamethasone for relapsed or refractory multiple myeloma. N Engl J Med 357:2123–2132
59. Mitsiades CS, Mitsiades NS, McMullan CJ et al (2006) Antimyeloma activity of heat shock protein-90 inhibition. Blood 107:1092–1100
60. Richardson PG, Chanan-Khan A, Lonial S et al (2009) Tanespimycin plus bortezomib in patients with relapsed and refractory multiple myeloma: final results of a phase I/II study. Abstract 8503. J Clin Oncol 27:434s

Bortezomib in Mantle Cell Lymphoma

Andre Goy

Abstract Since its recognition as a separate subtype of non-Hodgkin lymphoma (NHL) in 1994, mantle cell lymphoma (MCL) has been a very active field of clinical research due to its typical poor outcome. Besides a small subset of patients who can enjoy long-term disease-frees urvival through non-myeloablative allogenic stem cell transplant, most patients relapse and become frequently over time chemoresistant to conventional or high-dose cytotoxic-based therapy. Thankfully a number of novel agents have shown activity in MCL, most of which target distinct newly identified pathways. Proteasome inhibitors represent the first novel biological agent with promising activity in MCL with bortezomib as first in class and firstnovel agent FDA approved in relapsed/refractory MCL. Pre-clinical studies showing impressive additive or synergistic effects with other cytotoxics or biologicals, provide a rationale for a number of ongoing studies looking at combination with conventional regimens used in MCL either with chemoimmunotherapy or as consolidation or maintenance approaches post-induction. The progress in understanding MCL biology especially regarding the heterogeneity of the disease as well as emerging predictive biomarkers (both for conventional and novel therapies), will help stratify patients and refine our management to hopefully continue to improve patients' outcome.

1 Mantle Cell Lymphoma

Mantle cell lymphoma (MCL), a distinct subtype of B-cell lymphoma arising from the mantle zone of the lymph node follicle, accounts for approximately 6% of Non-Hodgkin Lymphoma (NHL) cases. MCL cells are derived from pre-germinal center

A. Goy
John Theurer Cancer Center at HUMC, Hackensack, NJ, USA
e-mail: agoy@humed.com

I.M. Ghobrial et al. (eds.), *Bortezomib in the Treatment of Multiple Myeloma*, Milestones in Drug Therapy,
DOI 10.1007/978-3-7643-8948-2_8,

cells and carry the immunophenotype of a mature B-cell (CD19, CD20) co-expressing CD5 but negative for CD23, CD10 and bcl-6; diagnosis of MCL requires demonstration of cyclin D1 overexpression by nuclear staining or the presence of the t(11;14)(q13;q32) translocation by fluorescence in situ hybridization (FISH). A small subset of cyclin-D1-negative MCL shows overexpression of cyclin D2 or D3, but otherwise, shares similar presentation and outcome with cyclin-D1-positive MCL [1, 2]. At diagnosis, patients with MCL are typically aged in their early-to-mid 60s with advanced-stage disease (>70% Ann Arbor stage IV disease) presenting with lymphadenopathy, splenomegaly, and almost constant extra-nodal involvement including bone marrow, liver, or gastrointestinal involvement (present in >90% cases at baseline), although B symptoms (fever, weight loss, or night sweats) are seen in less than one-third of the patients [3–6]. MCL circulating cells are present in most cases, but to varying degrees [7]. A prognostic index, the MCL International Prognostic Index (MIPI), which incorporates age, ECOG performance status, lactate dehydrogenase level, and leukocyte count, has recently been developed. Though modestly complex (calculation model based on continuous variables) it might help predict MCL patients' outcome [8]. Other prognostic factors in MCL include beta-2 microglobulin, presence of minimal residual disease post chemotherapy, various cytogenetic markers (such as 3q gains and 9q losses), p53 mutation/deletion (more common in blastoid variant) and proliferation signature [9–11]. Ki-67 (or MIB-1) has been used as a surrogate marker of the proliferation signature now validated in the rituximab-chemotherapy era. Among all lymphoma subtypes, MCL has one of the poorest prognoses with common chemoresistance over time. While median survival has improved over the past few decades as a result of multifactorial advances in disease management, median survival currently remains in the range of 4–5 years improved over the last few years (more than doubled) [12, 13].

2 Current Treatment of Mantle Cell Lymphoma

There is currently no consensus for management and no established curative approach for MCL. Despite initial response rates of 50–70% to many chemotherapy regimens, the typical pattern is relapse over time and development of chemoresistance. CHOP-based chemotherapy (cyclophosphamide, doxorubicin, vincristine, and prednisone) remains by default often the standard of care. The addition of rituximab to CHOP has improved overall and complete response rates (ORR and CR) and time to progression (TTP), but had no impact on progression-free survival (PFS) which remains at 25% at 2 years [14]. R-HyperCVAD (fractionated cyclophosphamide, vincristine, doxorubicin, and dexamethasone alternating with methotrexate and cytarabine) has been shown to be very promising as front-line therapy, with a more recent update at >9-year follow-up demonstrating a failure-free survival rate of 52% in a subgroup of patients aged <65 years [15, 16]. In an attempt to improve outcome, high-dose therapy (HDT) followed by autologous stem cell transplantation (ASCT) has also been studied extensively both in the front-line and relapsed-refractory setting [17, 18]. Although studies have clearly shown prolonged PFS with HDT followed by ASCT compared with standard-dose therapy, patients continue to relapse and there is no evidence that these new approaches are curative [19, 20]. With the aim of reducing hematological toxicities, a modified version of R-hyper-CVAD has been developed (no methotrexate or cytarabine, no vincristine on day 11, and no steroids on days 11–14) with added rituximab maintenance (four weekly doses every 6 months for 2 years); providing an ORR of 77%, CR of 64%, and median PFS of 37 months [21]. Nucleoside analogs have shown activity in MCL, especially fludarabine; a combination regimen of fludarabine, cyclophosphamide, and mitoxantrone with or without rituximab showed a clear benefit in the rituximab arm, with further benefit obtained from a short course (2 cycles at month 4 and 9) of rituximab maintenance therapy [22]. It has been hypothesized that a subset of more indolent MCL (5–10% of cases), with a mainly non-nodal presentation (mostly leukemic phase and splenomegaly and often a mutated somatic mutations pattern as seen in chronic lymphocytic leukemia), may qualify for a "watch and wait" approach, at least initially. A more detailed panorama of changing treatment options in MCL can be found in recent reviews [23, 24].

Following relapse, the median life expectancy for a patient with MCL declines to 1–2 years, and very few effective treatment options are then available, highlighting the need for novel treatment strategies. In addition, given the median age at diagnosis (mid 60s) there is a need for suitable treatments for this older patient group. Thanks to a growing number of novel biological agents, including bortezomib, the treatment paradigm of MCL is rapidly evolving.

3 Bortezomib in the Treatment of Mantle Cell Lymphoma

The proteasome is a large multiunit protease complex, which processes most intracellular proteins ($0.5–2 \times 10^6$ proteins per mn per cell) [25], including abnormal proteins (which are more abundant in cancer cells) and short-lived proteins that are typically involved in critical functions such as cell cycle, cell growth, differentiation and apoptosis [26]. Proteasome inhibition results in the disruption of a variety of pathways and checkpoints, which results in apoptosis. Bortezomib, first among the class of proteasome inhibitors, is a small molecule with potent selective and reversible inhibition of the chymotrypsine-like enzyme (one of the three enzymes located within the core of the proteasome) [27]. Bortezomib showed initial activity in a variety of lymphoma cell lines and in xenogaft models and early phase I studies showed activity at the highest bortezomib dose in two patients with NHL, including one patient with MCL [28]. Preclinical studies comparing responses to bortezomib among B-cell NHLs suggested that bortezomib was especially active in MCL, including in *ex vivo* cell samples from patients, and induced distinct changes in MCL compared to other types of lymphoma, including the overexpression of the anti-apoptotic protein Myeloid Cell Leukemia 1 (MCL-1) and induction of apoptosis through NOXA and Reactive Oxygen Species (ROS) [29], providing a rationale for further testing.

3.1 Clinical Efficacy of Single-Agent Bortezomib

The activity of single-agent bortezomib in MCL has been demonstrated in five phase 2 studies (Table 1) [30–35]. In these studies, patients received bortezomib 1.3 or 1.5 mg/m^2 on days 1, 4, 8, and 11 of a 21-day cycle. Unless treatment was discontinued for progressive disease or unacceptable toxicity, patients could receive 6–8 cycles in 3 of the studies [31–33] or until maximum response or 2 cycles beyond CR in one of the studies [30]. The PINNACLE trial, the largest study to date in relapsed or refractory MCL ($n = 141$), had a similar design using bortezomib 1.3 mg/m^2 bi-weekly for four cycles beyond CR or best response or up to 1 year (or 17 cycles) [35]. This pivotal trial showed an ORR of 33% with 8% CR/CRu [35], confirming, in the multicenter setting, the previously reported single institution phase II studies with an ORR of 29–46% and CR-CRu of 4–21% in relapsed–refractory MCL [30–34]. Noticeably MCL tends to respond faster than other types of NHL with a median time to response of about 6 weeks (first two cycles), including in patients with bulky disease. The median duration of response of 9–10 months was particularly impressive for a single agent in relapsed or refractory MCL. In the study by Goy et al., an estimated 80% of responders remained in remission at 6 months' and some patients remained progression-free for 2 years after treatment [32]. A recent update of the PINNACLE trial showed that the overall median time to progression was 6.7 months, while in responders, time to progression was 12.4 months [36]. Likewise, median overall survival was 23.5 months in all patients and 35.4 months among responders (Fig. 1). A recent update with subset analysis is summarized in Table 2. Patients with CR/CRu (half of whom had lesions >5 cm at study entry) had not reached median DOR at 27 months. Similarly, in the multicenter study by O'Connor et al. ($n = 37$) [30, 34], responding patients

Table 1 Phase 2 studies of single-agent bortezomib in MCL

Study	N	Dose	CR/CRu	PR	ORR
O'Connor et al. [30, 34]	40	1.5 mg/m^2	5 (12.5%)	14 (35%)	47%
Goy et al. [32]	29	1.5 mg/m^2	6 (20.5%)	6 (20.5%)	41%
Strauss et al. [31]	24	1.3 mg/m^2	1 (4%)	6 (25%)	29%
Belch et al. 2007 [33]	13 untreated	1.3 mg/m^2	0	6 (46%)	46%
	15 relapsed		1 (7%)	6 (40%)	47%
Fisher et al. [35] (PINNACLE)	141	1.3 mg/m^2	11 (8%)	36 (26%)	33%
Total	262		24 (9%)	74 (28%)	37%

CR complete response, *CRu* unconfirmed complete response, *PR* partial response, *ORR* overall response rate (CR + PR)

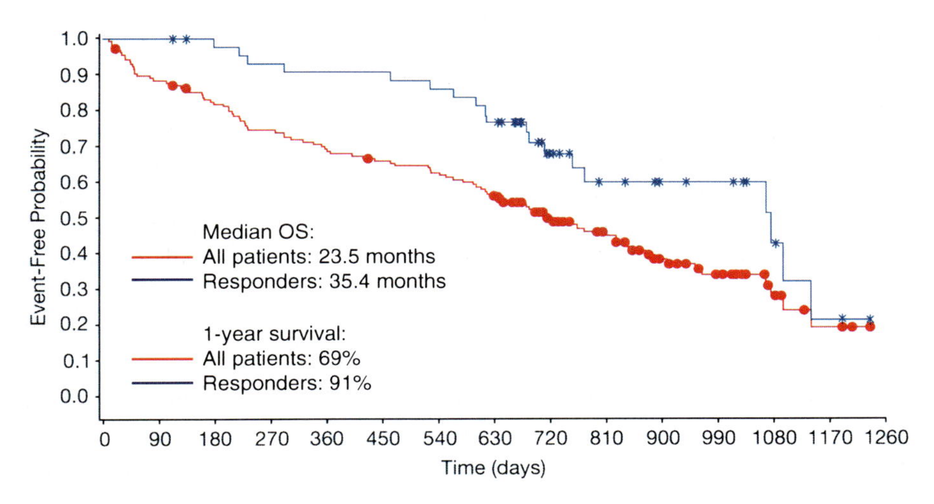

Fig. 1 Overall survival in all patients and in responders to bortezomib treatment in the PINNACLE study [36]

Table 2 Median time-to-event data in all patients, subgroups, and by response to bortezomib in the PINNACLE study [36]

Parameter (months)	All patients ($N = 155$)	Refractory MCL ($n = 58$)	Prior high-intensity therapy ($n = 58$)	Responders ($n = 45$)	CR/CRu ($n = 11$)	PR ($n = 34$)
TTP	6.7	3.9	4.2	12.4	NR	9.1
DOR	NA	5.9	NR	9.2	NR	6.7
OS	23.5	17.3	20.3	35.4	36.0	35.1

CR complete response, *CRu* unconfirmed complete response, *DOR* duration of response, *OS* overall survival, *PR* partial response, *TTP* time to progression NR (not reached)

demonstrated an event-free survival of about 10 months after bortezomib, compared to 6.2 months after their last therapy.

An important observation in these studies is the activity of bortezomib seen in patients who had received prior high-intensity therapy or in those with MCL refractory to previous therapy. For example, for the 37% of patients in the PINNACLE study who had received prior high-intensity therapy; the ORR was 25% and the median duration of response was not reached at publication [36]. In the study by Goy et al., a response rate of 41% was achieved despite 81% of patients having previously received the R-hyperCVAD regimen [32]. Similarly, in the original phase II studies, the number of prior therapies did not seem to affect response to bortezomib and responses were seen in the PINNACLE trial in both refractory and relapsed disease, suggesting that cross-resistance with conventional cytotoxic agents does not occur [34–36]. In addition, in the study by Goy et al., some patients achieved CR after failing 4–10 prior therapies [32]. There are also data suggesting that the duration of response with bortezomib in pretreated patients can be longer than that achieved with previous lines of chemotherapy [30]. This lack of cross-resistance with chemotherapeutic agents suggests different mechanisms of resistance to bortezomib and conventional cytotoxics; interestingly, bortezomib does not appear to be a good substrate for MDR. Furthermore, similar response rates were observed in untreated versus treated MCL in the study by Belch et al. [33]. The activity of bortezomib as a single agent though, seems to require a twice-weekly schedule. A recent study using a weekly schedule of bortezomib (up to 1.7 mg/m^2) in indolent lymphoma including MCL found that although the weekly bortezomib schedule was better tolerated than the twice-weekly schedule, the ORR was inferior (18 vs. 50%, $P = 0.02$) [37].

3.2 Safety and Tolerability Profile of Single-Agent Bortezomib

The toxicities seen with bortezomib in MCL (Table 3) are broadly similar to those seen in studies of bortezomib in multiple myeloma [38–41]. The most common adverse events included fatigue, peripheral neuropathy, hematological toxicities, and gastrointestinal events. In multiple myeloma studies, bortezomib-associated peripheral neuropathy and hematological toxicities (thrombocytopenia and neutropenia) have been shown to be reversible in most patients [39]. Notably, thrombocytopenia has been shown to be cyclical (inhibition of the NFKB-dependent mechanism of shedding megakarycotyes), recovering rapidly between treatment cycles [42].

In the MCL study by Belch et al., five patients had serious adverse events involving a combination of fluid retention and edema [33], resulting in a protocol amendment to exclude patients with such relevant pre-existing conditions; no further serious adverse events of this nature were subsequently reported. Skin rash occurred in approximately one-third of patients in the PINNACLE study (Fig. 2) [35], and while its mechanisms remains elusive, rash was more frequently reported with bortezomib treatment in MCL compared with other lymphomas. Unlike classic hypersensitivity type reactions, this vasculitic rash does not warrant cessation of the drug but can be and should be managed with supportive care

Table 3 Most common adverse events reported in ≥20% of patients receiving bortezomib in the PINNACLE study ($n = 155$) (adapted with permission from Fisher et al. [35])

Event	Any grade, *n* (%)	Grade ≥3, *n* (%)
Fatigue	95 (61%)	19 (12%)
Peripheral neuropathy	85 (55%)	20 (13%)
Constipation	77 (50%)	4 (3%)
Diarrhea	73 (47%)	11 (7%)
Rash	43 (28%)	4 (3%)
Vomiting	42 (27%)	4 (3%)
Anorexia	36 (23%)	5 (3%)
Dizziness (excluding vertigo)	36 (23%)	5 (3%)
Dyspnea	35 (23%)	5 (3%)
Insomnia	33 (21%)	1 (<1%)
Thrombocytopenia	33 (21%)	17 (11%)

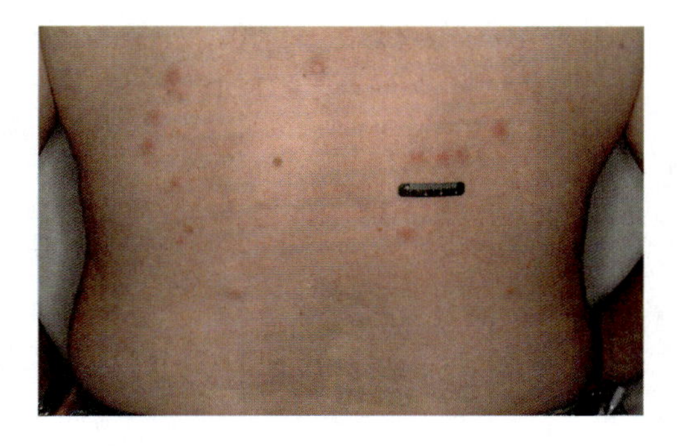

Fig. 2 Typical example of rash observed in phase II single-agent bortezomib studies

(such as topical corticosteroids or oral or topical anti-histamines), as it seems to be a surrogate marker predictive of response to bortezomib in MCL and NHL. In pooled data from 140 patients from phase II studies (among whom 26 developed a rash), the ORR was 73 vs. 33% in patients who did not develop rash [43]. Other toxicities included hypotension, fatigue, and malaise, which were improved by routine administration of intravenous fluids after bortezomib injection. Lymphopenia was also common and approximately 20% of patients develop herpes reactivation [44, 45], leading to antiviral routine prophylaxis recommendation.

Based on the efficacy and tolerability data detailed above, bortezomib became, in December 2006, the first drug approved by the FDA as second line in MCL for the treatment of relapsed or refractory disease.

3.3 Biomarkers

The similar response rates obtained with bortezomib (35–45%) across all studies underlines the need to define biomarkers that can predict response to bortezomib in MCL. Immunohistochemical analysis of samples from 70 patients in the PINNACLE study showed, as expected in this population, a correlation between Ki-67 level and disease progression [46]. In addition, higher expression of tumor α-5 proteasome subunit (which carries the chymotrysin-like enzymatic site) was associated with shorter time to progression, while higher levels of NF-κB p65 (a putative important target of bortezomib) were associated with longer time to progression. Similarly, in a tissue microarray analysis of samples from patients enrolled in the study originally reported by O'Connor et al. [30], NHL patients with pre-treatment positivity for p27 were more likely to respond to bortezomib and had a longer duration of response, while Bcl-6 appeared as a negative predictor (potentially reflecting the non-MCL subset in that trial) [47]. These preliminary findings suggest that distinct biological pathways, as seen in multiple myeloma, may be relevant to outcomes after bortezomib. Though the data remain at this point hypothesis generating, they warrant further evaluation for prognostic utility using routine built-in collection of materials at study entry in future studies.

3.4 Bortezomib Combination Studies

Additive or synergistic effects have been shown between bortezomib and a variety of other biologicals, conventional cytotoxics, and radiation, consistent with the biology of proteasome inhibition affecting multiple pathways including survival and apoptosis pathways [48, 49]. This provides a rationale for combination therapy (Table 4), with a rapidly growing number of combinations being explored both in the frontline as well as relapsed and refractory setting. Overall, more than 100 studies investigating bortezomib regimens in lymphomas are listed on http://clinicaltrials.gov.

In front-line MCL therapy, combinations of bortezomib with R-CHOP have demonstrated feasibility with promising preliminary efficacy [50, 51]. An ongoing phase 3 randomized study is comparing R-CHOP versus R-CHBzP (replacing vincristine with bortezomib in R-CHOP) in patients aged over 60 years. Phase I and II studies have demonstrated the efficacy of the combination of bortezomib and modified R-CVAD or R-HyperCVAD, with evidence that the addition of bortezomib may enhance efficacy compared with chemotherapy alone [52, 53]. For example, in one study, 30 newly diagnosed MCL patients received bortezomib on days 1 and 4 of 6 cycles of modified R-CVAD followed by maintenance treatment with rituximab [52]. After the addition of bortezomib, the ORR was 90% with 77% CR-CRu (compared with 60% with modified R-CVAD without bortezomib). As expected, the toxicities were primarily hematological, although the incidence of neuropathy prompted a dose reduction of bortezomib from 1.5 to 1.3 mg/m^2 and capping of vincristine at 1 mg. An ongoing CALGB study is comparing two regimens of bortezomib following HDT and ASCT: a maintenance regimen (bortezomib on days 1, 8, 15, and 22 every 56 days for up to ten cycles) and a consolidation regimen (bortezomib on days 1, 4, 8, and 11 every 21 days for up to four cycles). An NCI study using an R-EPOCH-bortezomib combination is exploring a window of opportunity as a single agent (with analysis of biomarkers on circulating MCL cells) as well as randomized maintenance with bortezomib post completion of therapy [54].

In relapsed or refractory MCL, combinations of bortezomib with chemotherapy and with biological agents are also being investigated. For example, one ongoing study is investigating R-CBoRP (replacement of vincristine with bortezomib in the R-CVP regimen) [55]. A combination with bortezomib, rituximab and dexamethasone has been reported (BoRID) with very promising activity [56]. A combination of bortezomib with fludarabine and rituximab has shown evidence of activity and acceptable tolerability in a recent Phase I study [57], and combinations of bortezomib with bendamustine or flavopiridol have also shown activity and acceptable tolerability in small studies involving patients with MCL [58, 59]. Preliminary promising activity has also been demonstrated with bortezomib alongside radioimmunotherapy [60].

Unfortunately, many patients ultimately develop resistance to single agent bortezomib. Mechanisms of resistance to proteasome inhibition appear to involve

Table 4 Bortezomib combination studies in MCL

Combination	Regimen	Results
Front-line MCL, CHOP backbone		
Ribrag et al. [51] (GELA)	Bortezomib + R-CHOP $n = 49$ ($n = 4$ with MCL)	Overall population: CR/CRu 41 patients (84%), PR 6 patients, neurotoxicity in 21 patients
Leonard et al. [50]	Bortezomib + R-CHOP $n = 20$ ($n = 4$ with MCL)	Overall population: CR/CRu 80%, ORR 95%, generally well tolerated
Clinicaltrials.gov ID NCT00376961 (SWOG)	Bortezomib + R-CHOP induction + 8 cycles of maintenance Estimated $n = 60$	Ongoing, primary outcome PFS at 2 years
Clinicaltrials.gov ID NCT00722137	R-CHOP vs. R-CHBzP (replacing vincristine with bortezomib in R-CHOP) Estimated $n = 486$	Ongoing, primary outcome PFS
Gressin et al. [71] (GOELAMS)	Bortezomib + rituximab + dexamethasone + doxorubicin + chlorambucil (RiPAD + C) Estimated $n = 27$, patients aged ≥ 65 years	10 patients in response (3 in CR, 7 in PR), grade 3–4 toxicities reported in 5 cases out of 59 cycles (2 neurologic, 1 hepatic, 1 cardiac, 1 pulmonary)
Frontline MCL, dose-dense or high-dose ASCT		
Kahl et al. [52]	Bortezomib + R-CVAD $n = 30$	CR/CRu 77%, PR 13%, toxicities mainly hematologic and peripheral neuropathies
Romaguera et al. [53]	Bortezomib + R-HyperCVAD/ methotrexate + cytarabine $n = 16$	CR 73%, PR 27%, mainly hematologic toxicities
Clinicaltrials.gov ID NCT00433537 (ECOG)	Bortezomib + R-CVAD with rituximab maintenance Estimated $n = 72$	Ongoing, primary endpoint CR
Clinicaltrials.gov ID NCT00310037 (CALGB)	Bortezomib after combination chemotherapy + rituximab + ASCT (maintenance vs consolidation therapy) Estimated $n = 98$	Ongoing, primary endpoint PFS at 18 months
Clinicaltrials.gov ID NCT00571493 (University of Nebraska)	Bortezomib + BEAM + HDT-ASCT Estimated $n = 44$	Ongoing, primary endpoint MTD
Clinicaltrials.gov ID NCT00131976 (NCI)	Bortezomib + EPOCH-R + randomized maintenance with rituximab Estimated $n = 80$	Ongoing, primary endpoints PFS and OS at 5 years
Relapsed MCL: combination with other cytotoxics		
Beavan et al. [60]	Bortezomib + RIT (^{90}yttrium-labelled ibritumomab tiuxetan) $n = 12$ ($n = 5$ with MCL)	2 MCL patients progression-free after 11 months
Clinicaltrial.gov ID NCT00777114	Bortezomib + RIT (^{131}I-tositumomab) Estimated $n = 25$	Ongoing phase I
Gerecitano et al. [72]	R-CBorP (replacing vincristine with bortezomib in R-CVP) $n = 42$ ($n = 9$ MCL)	Of 27 assessable patients, CR 5 patients, PR 13 patients, SD 8 patients.

(*continued*)

Table 4 (continued)

Combination	Regimen	Results
		Randomized phase II schedules (weekly versus bi-weekly) are ongoing.
Moosmann et al. [59]	Bortezomib + bendamustin $n = 9$ ($n = 4$ with MCL)	PR 4/4 MCL patients, acceptable tolerability
Weigert et al. [73]	Bortezomib 1.5 mg/m^2 on days 1 and 4 + Ara-C 750–2000 mg/m^2 on days 2–3 + dexamethasone + rituximab $n = 8$	ORR 4/8 patients, CR 2 patients, median PFS 5 months, median OS 15.5 months, grade 3/4 hematoxicity in all patients
Clinicaltrials.gov ID NCT00377052 (NCI)	Bortezomib + gemcitabine Estimated $n = 46$	Ongoing, primary endpoints ORR, duration of response, and TTP
Clinicaltrials.gov ID NCT00513955 (UK)	Bortezomib + CHOP Estimated $n = 90$	Ongoing, primary endpoint disease progression
Barr et al. [57]	Bortezomib + fludarabine + rituximab $n = 23$ ($n = 2$ with MCL)	Phase I study suggests regimen is active and well-tolerated
Clinicaltrials.gov ID NCT00295932 (Memorial Sloan-Kettering Cancer Center)	Bortezomib + rituximab + cyclophosphamide + prednisone (CBorP) Estimated $n = 115$	Ongoing, primary endpoint MDT
Clinicaltrials.gov ID NCT00711828 (Mayo Clinic)	Bortezomib + rituximab + cyclophosphamide + dexamethasone (R-CyBor-D) Estimated $n = 36$	Ongoing, primary endpoints CR and PR
Relapsed MCL: combination with other biologicals		
Drach et al. [56] (BORID Trial)	Bortezomib + dexamethasone + rituximab ($n = 10$)	CR 3 patients, PR 4 patients, 6/6 patients progression free at 6 months, toxicities mainly neuropathy and thrombocytopenia
Clinicaltrials.gov ID NCT00703664 (H. Lee Moffitt Cancer Center)	Bortezomib + vorinostat Estimated $n = 116$	Ongoing, primary endpoint ORR
Clinicaltrials.gov ID NCT00791011	Bortezomib + vorinostat + AMG655 Estimated $n = 62$	Ongoing, primary endpoints safety, ORR
Clinicaltrials.gov ID NCT00671112 (NCI)	Bortezomib + everolimus Estimated $n = 36$	Ongoing, primary endpoint MTD
Clinicaltrials.gov ID NCT00407303	Bortezomib + obatoclax Estimated $n = 60$	Ongoing, primary endpoint ORR (out of 16 evaluable patients, 3 CR, 2 PR, 5 SD)
Grant et al. [58]	Bortezomib + flavopiridol $n = 38$	CR 2 patients (including 1 MCL patient), generally well tolerated, MTD not reached
Clinicaltrials.gov ID NCT00553644 (CALGB)	Bortezomib + lenalidomide Estimated $n = 54$	Ongoing, primary endpoint ORR
Clinicaltrials.gov ID NCT00633594 (Sarah Cannon Research Institute)	Bortezomib + rituximab + lenalidomide Estimated $n = 58$	Ongoing dose escalation study to determine MTD

CR/*CRu* complete response/unconfirmed complete response, *EFS* event-free survival, *MTD* maximum tolerated dose, *ORR* overall response rate, *PFS* progression-free survival, *PR* partial response, *RIT* radioimmunotherapy, *TTP* time to progression several factors, including those related to the proteasome itself, such as overexpression of the proteasome subunits or binding mutations, and factors outside of the proteasome, such as the accumulation of proteins upstream of the inhibited proteasome resulting in alternate pathways of protein degradation through the aggresome [61]. One important mechanism of resistance to proteasome inhibition appears to involve heat-shock proteins (HSP), especially HSP90 [62, 63]. HSPs have anti-apoptotic activity, and so induction of HSPs counteracts the proapoptotic activities of proteasome inhibition, thus leading to resistance [64, 65]. These observations provide a rationale for using bortezomib in combination with HSP inhibitors, some of which, such as the HSP90 inhibitor 17AGG (geldamycin), have shown preliminary evidence of activity in lymphoma on their own and are already used in combination regimens in the treatment of multiple myeloma in the clinic [66]. Histone deacetylase (HDAC) inhibitors have shown synergistic antitumor effects with bortezomib in preclinical models, supporting also ongoing trials [67, 68]. An impressive number of new combinations are being looked at in both front-line and relapsed MCL as summarized in Table 4. Of note, preclinical observations suggest that some of these combinations might be schedule-dependent or sequence-dependent, potentially influencing the design of future clinical trials [48, 49].

4 Second Generation Proteasome Inhibitors

Second generation proteasome inhibitors include NPI-0052 (salinosporin A, a marine derived proteasome inhibitor from algae species actinomycete Salinospora Tropica), which can inhibit all three enzymes of the 20S unit of the proteasome and has just entered the clinic, and PR-171 or carfilzomib, a novel, irreversible proteasome inhibitor under investigation in phase 1–2 studies. In spite of being irreversible, preclinical data showed that the half-life of recovery from proteasome inhibition by carfilzomib in tissues was approximately 24 h. Carfilzomib showed evidence of greater potency compared with bortezomib in preclinical models and is currently being tested with so far anecdotal responses in NHL including one MCL patient [69]. The next generation of proteasome pathway inhibitors includes ligase-specific inhibitors (to increase specificity) as well as immunoproteasome inhibitors that might increase the therapeutic index by preferentially targeting lymphoid cells where the ratio of immunoproteasomes to constitutive proteasome is higher [70].

5 Conclusions

Though a rare subtype of NHL, MCL represents a prime example of the ongoing changes in the field with many emerging novel therapies. Proteasome inhibitors, with bortezomib as first in this class, represent the first novel option approved by the

FDA in MCL. Ongoing and future studies will define the role of biologicals in the management of patients in combination with chemotherapy agents as well as the potential role of bortezomib as maintenance or consolidation post-induction, to hopefully continue to improve patients' outcome.

References

1. Herens C, Lambert F, Quintanilla-Martinez L et al (2008) Cyclin D1-negative mantle cell lymphoma with cryptic t(12;14)(p13;q32) and cyclin D2 overexpression. Blood 111:1745–1746
2. Wlodarska I, Dierickx D, Vanhentenrijk V et al (2008) Translocations targeting CCND2, CCND3, and MYCN do occur in t(11;14)-negative mantle cell lymphomas. Blood 111:5683–5690
3. Armitage JO (1998) Management of mantle cell lymphoma. Oncology (Williston Park) 12:49–55
4. Fisher RI (1996) Mantle-cell lymphoma: classification and therapeutic implications. Ann Oncol 7(Suppl 6):S35–S39
5. Argatoff LH, Connors JM, Klasa RJ et al (1997) Mantle cell lymphoma: a clinicopathologic study of 80 cases. Blood 89:2067–2078
6. Romaguera JE, Medeiros LJ, Hagemeister FB et al (2003) Frequency of gastrointestinal involvement and its clinical significance in mantle cell lymphoma. Cancer 97:586–591
7. Cohen PL, Kurtin PJ, Donovan KA, Hanson CA (1998) Bone marrow and peripheral blood involvement in mantle cell lymphoma. Br J Haematol 101:302–310
8. Salek S, Vasova I, Pytlik R et al (2008) Mantle cell lymphoma international prognostic score is valid and confirmed in unselected cohort of patients treated in rituximab era. Blood 112:3745
9. Pott C, Hoster E, Böttcher S et al (2008) Molecular remission after combined immunochemotherapy is of prognostic relevance in patients with MCL: results of the randomized intergroup trials of the European MCL Network. Blood 112:582
10. Salaverria I, Zettl A, Beà S et al (2007) Specific secondary genetic alterations in mantle cell lymphoma provide prognostic information independent of the gene expression-based proliferation signature. J Clin Oncol 25:1216–1222
11. Rosenwald A, Wright G, Wiestner A et al (2003) The proliferation gene expression signature is a quantitative integrator of oncogenic events that predicts survival in mantle cell lymphoma. Cancer Cell 3:185–197
12. Herrmann A, Hoster E, Zwingers T et al (2009) Improvement of overall survival in advanced stage mantle cell lymphoma. J Clin Oncol 27:511–518
13. Lenz G, Dreyling M, Hiddemann W (2004) Mantle cell lymphoma: established therapeutic options and future directions. Ann Hematol 83:71–77
14. Howard OM, Gribben JG, Neuberg DS et al (2002) Rituximab and CHOP induction therapy for newly diagnosed mantle-cell lymphoma: molecular complete responses are not predictive of progression-free survival. J Clin Oncol 20:1288–1294
15. Romaguera JE, Fayad L, Rodriguez MA et al (2005) High rate of durable remissions after treatment of newly diagnosed aggressive mantle-cell lymphoma with rituximab plus hyper-CVAD alternating with rituximab plus high-dose methotrexate and cytarabine. J Clin Oncol 23:7013–7023
16. Romaguera R, Fayad L, Rodriguez A et al (2008) Rituximab (R) + HyperCVAD alternating with R-methotrexate/cytarabine after 9 years: continued high rate of failure-free survival in untreated mantle cell lymphoma (MCL). Blood 112:833
17. Jacobsen E, Freedman A (2004) An update on the role of high-dose therapy with autologous or allogeneic stem cell transplantation in mantle cell lymphoma. Curr Opin Oncol 16:106–113

18. Sweetenham JW (2001) Stem cell transplantation for mantle cell lymphoma: should it ever be used outside clinical trials? Bone Marrow Transplant 28:813–820
19. Kasamon YL (2007) Blood or marrow transplantation for mantle cell lymphoma. Curr Opin Oncol 19:128–135
20. Sweetenham JW (2009) Review: stem cell transplantation for mantle cell lymphoma: not yet the standard of care. Clin Adv Hematol Oncol 7:323–324
21. Kahl BS, Longo WL, Eickhoff JC et al (2006) Maintenance rituximab following induction chemoimmunotherapy may prolong progression-free survival in mantle cell lymphoma: a pilot study from the Wisconsin Oncology Network. Ann Oncol 17:1418–1423
22. Forstpointner R, Unterhalt M, Dreyling M et al (2006) Maintenance therapy with rituximab leads to a significant prolongation of response duration after salvage therapy with a combination of rituximab, fludarabine, cyclophosphamide, and mitoxantrone (R-FCM) in patients with recurring and refractory follicular and mantle cell lymphomas: results of a prospective randomized study of the German Low Grade Lymphoma Study Group (GLSG). Blood 108:4003–4008
23. Goy A (2007) Mantle cell lymphoma: evolving novel options. Curr Oncol Rep 9:391–398
24. Goy A, Feldman T (2007) Expanding therapeutic options in mantle cell lymphoma. Clin Lymphoma Myeloma 7(Suppl 5):S184–S191
25. Yewdell JW, Reits E, Neefjes J (2003) Making sense of mass destruction: quantitating MHC class I antigen presentation. Nat Rev Immunol 3:952–961
26. Adams J (2004) The proteasome: a suitable antineoplastic target. Nat Rev Cancer 4:349–360
27. Adams J, Palombella VJ, Sausville EA et al (1999) Proteasome inhibitors: a novel class of potent and effective antitumor agents. Cancer Res 59:2615–2622
28. Orlowski RZ, Stinchcombe TE, Mitchell BS et al (2002) Phase I trial of the proteasome inhibitor PS-341 in patients with refractory hematologic malignancies. J Clin Oncol 20:4420–4427
29. Pérez-Galán P, Roué G, Villamor N et al (2006) The proteasome inhibitor bortezomib induces apoptosis in mantle-cell lymphoma through generation of ROS and Noxa activation independent of p53 status. Blood 107:257–264
30. O'Connor OA, Wright J, Moskowitz C et al (2005) Phase II clinical experience with the novel proteasome inhibitor bortezomib in patients with indolent non-Hodgkin's lymphoma and mantle cell lymphoma. J Clin Oncol 23:676–684
31. Strauss SJ, Maharaj L, Hoare S et al (2006) Bortezomib therapy in patients with relapsed or refractory lymphoma: potential correlation of in vitro sensitivity and tumor necrosis factor alpha response with clinical activity. J Clin Oncol 24:2105–2112
32. Goy A, Younes A, McLaughlin P et al (2005) Phase II study of proteasome inhibitor bortezomib in relapsed or refractory B-cell non-Hodgkin's lymphoma. J Clin Oncol 23:667–675
33. Belch A, Kouroukis CT, Crump M et al (2007) A phase II study of bortezomib in mantle cell lymphoma: the National Cancer Institute of Canada Clinical Trials Group trial IND.150. Ann Oncol 18:116–121
34. O'Connor OA, Moskowitz C, Portlock C et al (2009) Patients with chemotherapy-refractory mantle cell lymphoma experience high response rates and identical progression-free survivals compared with patients with relapsed disease following treatment with single agent bortezomib: results of a multicentre phase 2 clinical trial. Br J Haematol 145:34–39
35. Fisher RI, Bernstein SH, Kahl BS et al (2006) Multicenter phase II study of bortezomib in patients with relapsed or refractory mantle cell lymphoma. J Clin Oncol 24:4867–4874
36. Goy A, Bernstein SH, Kahl BS et al (2009) Bortezomib in patients with relapsed or refractory mantle cell lymphoma: updated time-to-event analyses of the multicenter phase 2 PINNACLE study. Ann Oncol 20:520–525
37. Gerecitano J, Portlock C, Moskowitz C et al (2009) Phase 2 study of weekly bortezomib in mantle cell and follicular lymphoma. Br J Haematol 0146(6):652–655, Jul 16
38. Jagannath S, Barlogie B, Berenson JR et al (2005) Bortezomib in recurrent and/or refractory multiple myeloma. Initial clinical experience in patients with impaired renal function. Cancer 103:1195–1200

39. Richardson PG, Briemberg H, Jagannath S et al (2006) Frequency, characteristics, and reversibility of peripheral neuropathy during treatment of advanced multiple myeloma with bortezomib. J Clin Oncol 24:3113–3120
40. Richardson PG, Sonneveld P, Schuster MW et al (2007) Safety and efficacy of bortezomib in high-risk and elderly patients with relapsed multiple myeloma. Br J Haematol 137:429–435
41. Berenson JR, Jagannath S, Barlogie B et al (2005) Safety of prolonged therapy with bortezomib in relapsed or refractory multiple myeloma. Cancer 104:2141–2148
42. Lonial S, Waller EK, Richardson PG et al (2005) Risk factors and kinetics of thrombocytopenia associated with bortezomib for relapsed, refractory multiple myeloma. Blood 106:3777–3784
43. Gerecitano J, Goy A, Wright J et al (2006) Drug-induced cutaneous vasculitis in patients with non-Hodgkin lymphoma treated with the novel proteasome inhibitor bortezomib: a possible surrogate marker of response? Br J Haematol 134:391–398
44. Chanan-Khan A, Sonneveld P, Schuster MW et al (2008) Analysis of herpes zoster events among bortezomib-treated patients in the phase III APEX study. J Clin Oncol 26:4784–4790
45. Kim SJ, Kim K, Kim BS et al (2008) Bortezomib and the increased incidence of herpes zoster in patients with multiple myeloma. Clin Lymphoma Myeloma 8:237–240
46. Goy A, Bernstein SH, McDonald A et al (2007) Immunohistochemical analyses for potential biomarkers of bortezomib activity in mantle cell lymphoma from the PINNACLE phase 2 trial. Blood 110:2573
47. Gerecitano J, Gounder S, Feldstein J et al (2007) Pre-Treatment p27 and Bcl-6 staining levels correlate with response to bortezomib in non-Hodgkin lymphoma: Results from a tissue microarray analysis. Blood 110:1294
48. Goy A, Remache Y, Barkoh B et al (2004) Sensitivity, schedule-dependence and molecular effects of the proteasome inhibitor bortezomib in non-Hodgkin's lymphoma cells. Blood 104:389a
49. Weigert O, Pastore A, Rieken M et al (2007) Sequence-dependent synergy of the proteasome inhibitor bortezomib and cytarabine in mantle cell lymphoma. Leukemia 21:524–528
50. Leonard J, Furman R, Feldman E et al (2005) Phase I/II trial of bortezomib + CHOP-rituximab in diffuse large B cell (DLBCL) and mantle cell lymphoma (MCL): phase I results. Blood 104:491
51. Ribrag V, Gisselbrecht C, Haioun C et al (2009) Efficacy and toxicity of 2 schedules of frontline rituximab plus cyclophosphamide, doxorubicin, vincristine, and prednisone plus bortezomib in patients with B-cell lymphoma: a randomized phase 2 trial from the French Adult Lymphoma Study Group (GELA). Cancer 115(19):4540–4546, Jul 10
52. Kahl B, Chang J, Eickhoff J et al (2008) VcR-CVAD produces a high complete response rate in untreated mantle cell lymphoma: a phase II study from the Wisconsin Oncology Network. Blood 112:265
53. Romaguera J, Fayad L, McLaughlin P et al (2008) Phase I trial of bortezomib in combination with rituximab-HyperCVAD/methotrexate and cytarabine for untreated mantle cell lymphoma. Blood 112:3051
54. Wiestner A, Dunleavy K, Rizzatti EG et al (2005) Potent single agent activity and tumor selectivity of bortezomib in mantle cell lymphoma: first impressions from a randomized phase ii study of EPOCH-rituximab-bortezomib in untreated mantle cell lymphoma. Blood 106:4744
55. Gerecitano JF, Portlock C, Hamlin P et al (2008) A phase I study evaluating two dosing schedules of bortezomib (Bor) with rituximab I, cyclophosphamide I and prednisone (P) in patients with relapsed/refractory indolent and mantle cell lymphomas. J Clin Oncol 26(Suppl May 20):8512
56. Drach J, Kaufmann H, Pichelmayer O et al (2006) Marked activity of bortezomib, rituximab, and dexamethasone (BORID) in heavily pretreated patients with mantle cell lymphoma. J Clin Oncol 24(Suppl 18):17522
57. Barr PM, Fu P, Lazarus HM et al (2008) Phase I dose escalation study of fludarabine, bortezomib, and rituximab for relapsed/refractory indolent and mantle cell non-Hodgkin lymphoma. J Clin Oncol 26(Suppl):8553

58. Grant S, Sullivan D, Roodman D et al (2008) Phase I trial of bortezomib (NSC 681239) and flavopiridol (NSC 649890) in patients with recurrent or refractory indolent B-cell neoplasms. Blood 112:1573
59. Moosmann PR, Heizmann M, Kotrubczik N et al (2008) Weekly treatment with a combination of bendamustine and bortezomib in relapsed or refractory indolent non-hodgkin's lymphoma: a single-center phase 1/2 study. Blood 112:1574
60. Beaven A, Shea TC, Moore DT et al (2008) A Phase I study of bortezomib (Velcade®) plus ^{90}yttrium labeled ibritumomab tiuxetan (Zevalin®) in patients with relapsed or refractory B-cell non-hodgkin lymphoma (NHL). Blood 112:4944
61. Simms-Waldrip T, Rodriguez-Gonzalez A, Lin T et al (2008) The aggresome pathway as a target for therapy in hematologic malignancies. Mol Genet Metab 94:283–286
62. Mitsiades CS, Mitsiades NS, McMullan CJ et al (2006) Antimyeloma activity of heat shock protein-90 inhibition. Blood 107:1092–1100
63. Shringarpure R, Catley L, Bhole D et al (2006) Gene expression analysis of B-lymphoma cells resistant and sensitive to bortezomib. Br J Haematol 134:145–156
64. Chauhan D, Li G, Shringarpure R et al (2003) Blockade of hsp27 overcomes bortezomib/ proteasome inhibitor PS-341 resistance in lymphoma cells. Cancer Res 63:6174–6177
65. Gabai VL, Budagova KR, Sherman MY (2005) Increased expression of the major heat shock protein Hsp72 in human prostate carcinoma cells is dispensable for their viability but confers resistance to a variety of anticancer agents. Oncogene 24:3328–3338
66. Thomas X, Campos L, Le QH, Guyotat D (2005) Heat shock proteins and acute leukemias. Hematology 10:225–235
67. Mitsiades CS, Mitsiades NS, McMullan CJ et al (2004) Transcriptional signature of histone deacetylase inhibition in multiple myeloma: biological and clinical implications. Proc Natl Acad Sci USA 101:540–545
68. Pei XY, Dai Y, Grant S (2004) Synergistic induction of oxidative injury and apoptosis in human multiple myeloma cells by the proteasome inhibitor bortezomib and histone deacetylase inhibitors. Clin Cancer Res 10:3839–3852
69. Stewart KA, O'Connor OA, Alsina M et al (2007) Phase I evaluation of carfilzomib (PR-171) in haematological malignancies: responses in multiple myeloma and Waldenstrom's macroglobulinemia at well-tolerated doses. J Clin Oncol 25(Suppl 18):8003
70. Yewdell JW (2005) Immunoproteasomes: regulating the regulator. Proc Natl Acad Sci USA 102:9089–9090
71. Gressin R, Maugendre SC, Le Gouill S et al (2008) Interim results of the RiPAD+C regimen including Velcade in front line therapy for elderly patients with mantle cell lymphoma. a phase II prospective study of the GOELAMS Group. Blood 112:1575
72. Gerecitano JF, Portlock C, Hamlin P et al (2008) A phase I study evaluating two dosing schedules of bortezomib (Bor) with rituximab I, cyclophosphamide I and prednisone (P) in patients with relapsed/refractory indolent and mantle cell lymphomas. J Clin Oncol 26 (Suppl):8512
73. Weigert O, Weidmann E, Mueck R et al (2009) A novel regimen combining high dose cytarabine and bortezomib has activity in multiply relapsed and refractory mantle cell lymphoma – long-term results of a multicenter observation study. Leuk Lymphoma 50:716–722

Bortezomib in Waldenstrom's Macroglobulinemia

Irene M. Ghobrial, Aldo M. Roccaro, and Xavier Leleu

Abstract Waldenstrom's macroglobulinemia (WM) is a low-grade lymphoproliferative disorder characterized by the presence of a lymphoplasmacytic infiltrate in the bone marrow and the presence of a serum monoclonal protein IgM. Despite the clinical efficacy of conventional therapies, most patients eventually relapse and the disease remains incurable. Therefore, novel therapeutic agents are needed for the treatment of WM. The multicatalytic ubiquitin–proteasome pathway plays an important role in the targeted degradation of a wide spectrum of proteins involved in the regulation of several cellular processes. The proteasome itself has been selected as a target for cancer therapy. Bortezomib represents the first proteasome inhibitor to enter clinical trials. The enthusiastic preclinical and clinical results exerted by bortezomib in multiple myeloma, as well as other hematological malignancies including WM, has validated the idea that the proteasome is an important target in cancer therapy.

1 Introduction

Waldenstrom's macroglobulinemia (WM) represents a rare proliferative disorder characterized by accumulation of lymphoplasmacytic cells in the bone marrow accompanied by secretion of monoclonal immunoglobulin M (IgM) protein in the serum [1], which is considered as lymphoplasmacytic lymphoma, according to the Revised European American Lymphoma (REAL) and World Health Organization (WHO) classification systems [2]. In 1944, it was described for the first time by Jan Gosta Waldenstrom. WM has an overall incidence of approximately 3 per million

I.M. Ghobrial and A.M. Roccaro (✉)
Dana-Farber Cancer Institute, Medical Oncology, Harvard Medical School Boston, MA, USA
e-mail: aldo_roccaro@dfci.harvard.edu

X. Leleu
Department of Hematology, Service des Maladies du Sang, Hopital Huriez, Lille, France

I.M. Ghobrial et al. (eds.), *Bortezomib in the Treatment of Multiple Myeloma*,
Milestones in Drug Therapy,
DOI 10.1007/978-3-7643-8948-2_9,

persons per year, with almost 1,500 new cases diagnosed per year in the US [3, 4]. WM represents a sporadic disease and its cause remains unknown, although various reports indicate a high familial incidence of WM, along with any other B cell clonal disorder in first-degree relatives [5, 6]. It has been demonstrated that the main risk factor for the development of WM is the preexisting IgM-monoclonal gammopathy of undetermined significance (IgM-MGUS), which confers a 46-fold higher relative risk to develop WM than for the general population [7]. The origin of the malignant clone is thought to be a B cell that has been arrested after somatic hypermutation in the germinal center, before terminal differentiation to plasma cells [8, 9]. The malignant cells have undergone *VH* gene mutation, but not isotype class switching. Deletions in 6q21-22.1 were confirmed in most WM patients regardless of their family history [10], while analysis of translocations involving site 14q32, a common feature of many B cell malignancies, indicates the absence of Ig heavy chain (IgH) rearrangements in WM [11]. Many genes are thought to be dysregulated in WM but further studies to define the role of these genes in the pathogenesis of WM are underway [12].

Therapeutic options for WM include alkylating agents, nucleoside analogs, and rituximab as a single agent or in combination with other compounds. However, these approaches result in low complete response rates and short treatment-free survival intervals in most instances. Furthermore, no specific agent or regimen has been proven to be superior to others and, moreover, no treatment has been specifically approved for WM. As such, new approaches for the treatment of WM are needed. In an effort to achieve this, investigators have pursued targeted therapies such as bortezomib to exploit the biological characteristics of WM. On the basis of its activity in MM, bortezomib has been used in WM patients either as single agent or in combination with other drugs [13–16].

2 Bortezomib: Mechanism of Action

The antitumor activity exerted by bortezomib is the result of several actions mediated by the drug, such as NF-kB inhibition, upregulation of different proapoptotic pathways, downregulation of growth factor expression, and recently it has been demonstrated that bortezomib has anti-angiogenic activity (Fig. 1). Studies have shown that bortezomib targets NF-kB through its recruitment to the promoter of the target gene *IkB*, using chromatin immunoprecipitation assay in WM cells [17]. Bortezomib led to cytotoxicity in WM cells; this was mediated through a reduction of the PI3K/Akt signaling pathways, which is found to be critical for the survival of WM cells. By inhibiting NF-kB-mediated adhesion and cytokine secretion, bortezomib disrupts the main ways of tumor cell activation. Bortezomib is a proteasome inhibitor that induces apoptosis of primary WM lymphoplasmacytic cells, as well as WM cell lines, BCWM.1, and WSU-WM at pharmacologically achievable levels [17, 18]; moreover, bortezomib might also impact the bone

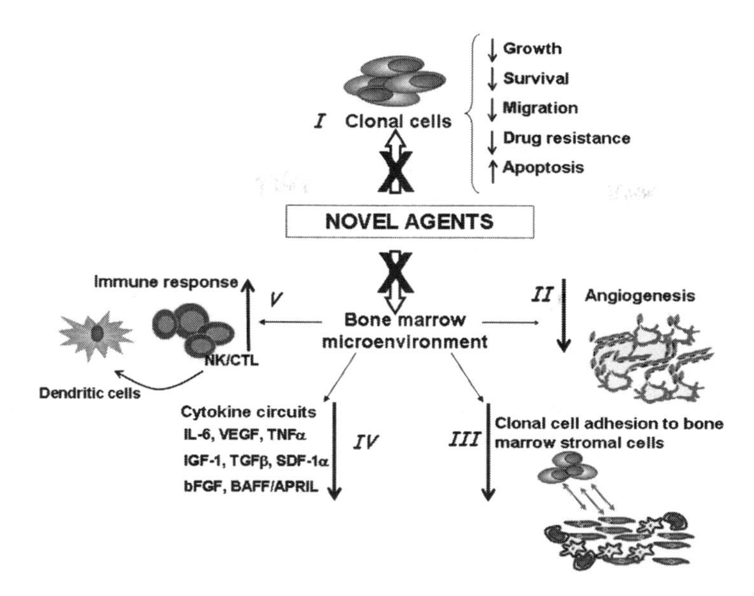

Fig. 1 Mechanism of action of Bortezomib in plasma cell malignancies: (I) Inhibition of plasma cell growth; (II) Reduced production of cytokines from bone marrow stromal cells; (III) Reduced expression of molecules involved in the interaction between plasma cells and bone marrow stromal cells; (IV) Anti-angiogenic effect

marrow microenvironment, which supports lymphoplasmacytic cell growth survival and resistance to treatment. Proteasome inhibitors have, therefore, become the focus of clinical research in WM.

2.1 *NF-kB Inhibition*

We have previously reported that NF-kB plays a crucial role in regulating WM pathogenesis and progression, providing the rational for using bortezomib in this disease. It has been clearly demonstrated that the inhibition of NF-kB, the major transcription factor belonging to the Rel/NF-kB family, mediates several biologic activities exerted by bortezomib [19]. It is a p50/RelA heterodimer (p50/p65) present in the cytoplasm of cells [20, 21]. NF-kB regulates cell growth and apoptosis, as well as the expression and the production of cytokines, adhesion molecules, and relative receptors [13]. In basal conditions, NF-kB is bound to its inhibitor I-kB [14]. Several factors such as cytokines, stress, or chemotherapeuticals may activate I-kB kinase, a heterodimeric protein kinase that phosphorylates two serine residues in the amino-terminal domain of I-kB [13]. The phosphorylated sites on I-kB are recognized by the ubiquitin ligase E3RS(I-kB/β-TrCP), leading to ubiquitination [15]. After degradation of the ubiquitinated I-kB by the proteasome pathway, NF-kB will be released in its free active form, which translocates to the nucleus where it binds to the promoter regions of different target genes and upregulates their transcription that finally leads to upregulated expression of cyclins, cytokines, and adhesion molecules that favor cell growth and survival [5]. Proteasome inhibitors have an inhibitory effect on NF-kB through a downregulation of the degradation of I-kB [14, 19, 22].

2.2 Induction of Proapoptotic Pathways

Several evidences show that bortezomib upregulates the expression of proapoptotic genes and conversely downregulates the expression of genes involved in proapoptotic pathways [23]. It has been demonstrated that cytotoxicity exerted by bortezomib in different tumor cell lines is mediated by apoptosis via activation of caspase-8 and/or -9, and -3, followed by PARP cleavage, despite the accumulation of p21Cip1 and p27Kip1 and irrespective of the p53 wild type or mutant status [23–26]. In particular, bortezomib induces apoptosis in drug-resistant multiple myeloma cells [24], and interleukin-6 (IL-6) does not overcome bortezomib-induced apoptosis in MM. Preclinical studies have further delineated the molecular mechanisms whereby bortezomib induces MM cell apoptosis [27, 28] and similar results were demonstrated in WM. The authors hypothesized that DNA damage triggered by bortezomib treatment in MM cells induces the activation of DNA-dependent protein kinase catalytic subunit (DNA-PKcs) and the activation of p53. They reported that the proteasome inhibitor induces p53 phosphorylation, which is associated with increased p53 protein expression. Bortezomib induces MDM2 protein and the association of p53 and MDM2 earlier than p53 phosphorylation, and these data demonstrate that p53 is still associated with MDM2 even after phosphorylation. Gene microarray data of MM cell lines treated with bortezomib show transcriptional triggering of apoptotic cascades, downregulation of growth and survival kinases, and induction of stress kinases including heat shock proteins. c-Jun NH2-terminal kinase (JNK) represents one of these stress-response proteins that mediates apoptosis triggered by unfolded proteins [29].

2.3 Inhibition of Angiogenic Phenotype

Proteasome inhibitors have greater activity against dividing endothelial cells than against quiescent cells [30], suggesting that they target the aberrant blood vessel development associated with tumor growth.

It has been demonstrated that lactacystin inhibits the in vitro capillarogenesis on Matrigel and also reduces the expression of plasminogen activator, an important protease involved in the angiogenic cascade [31]. Both lactacystin and the peptide aldehyde inhibitor PSI inhibit angiogenesis in the in vivo model of chick embryo chorioallantoic membrane (CAM) [30, 31], by inhibiting the formation of capillary vessels blood vessels of first-order [32].

On the basis of these findings, different authors have investigated the anti-angiogenic activity exerted in vivo by bortezomib in a murine model of MM, as well as of non-small cell lung carcinoma (NSCLC). The results demonstrate the following: (a) a significant reduction of tumor-related vascular density and (b) a downregulation of proangiogenic cytokines expression [33]. Recently, the anti-antiangiogenic activity exerted by bortezomib through its direct and indirect effect on endothelial cells isolated from bone marrow of patients with multiple myeloma (MMECs) has been investigated [34]. These data suggest that bortezomib, at clinically achievable concentrations, inhibits the proliferation of MMECs in a dose- and time-dependent manner. In functional assays of angiogenesis including chemotaxis, adhesion to fibronectin, capillary formation on matrigel, and CAM assay, bortezomib demonstrated a dose-dependent inhibition of angiogenesis. Importantly, binding of MM.1S cells to MMECs triggered-MM cell proliferation, which was also abrogated by bortezomib in a dose-dependent fashion. Dose-dependent downregulation of vascular endothelial growth factor (VEGF) and IL-6 secretion by MMECs was observed; furthermore, RT-PCR confirmed down-regulation of VEGF, IL-6, insulin-like growth factor-1 (IGF-1), as well as angio-poietin1 (Ang1) and Ang2 transcription in treated cells. These data demonstrate the anti-angiogenic effects of bortezomib, suggesting additional mechanisms for its activity against MM cells in the bone marrow milieu.

3 Clinical Studies

3.1 Bortezomib Single Agent Therapy in Waldenstrom's Macroglobulinemia

A retrospective analysis of ten WM patients who were previously treated with single agent bortezomib [13] administered intravenously (IV) with a dosage of 1.3 mg/m^2 on days 1, 4, 8, and 11 in a 21-day cycle for a total of 6 cycles, reported in 2005 six partial responses (PR), which occurred at a median time of 1 month. Bortezomib was relatively well tolerated and the median time to progression in the responding patients exceeded a year. On the basis of bortezomib activity in mye-loma and the encouraging results of this retrospective analysis, two phase II studies were conducted in relapse/refractory WM patients to further determine the toler-ance and the efficacy of bortezomib in WM disease.

The first phase II trial of single agent bortezomib was reported by the National Cancer Institute of Canada Clinical Trials group [9] in 27 symptomatic WM patients either untreated (44%) or relapsed (56%) disease. Bortezomib was admi-nistered I.V. at a dose of 1.3 mg/m^2 on days 1, 4, 8, and 11 in a 21-day cycle until progressive disease or until two cycles after complete response (CR). A median of six cycles (range: 2–39 cycles) were administered; 85% of patients received at least four cycles, and 33% received at least eight cycles. Twelve (44%) patients reached at least 50% using IgM reduction criteria, solely. However, when both IgM and bidimensional CT scan criteria applied, seven (26%) patients attained a major response (all PR). Response rates were similar between patients who had no prior therapy (25%) and those previously treated (27%). Of the 27 patients enrolled, 12 (44.4%) were discontinued from study for toxicity, most commonly because of neuropathy, and one (3.7%) for multifactorial debilitation. Dose reductions were required in eight patients, most commonly because of nonhematologic toxicity (neuropathy in four patients; various toxicities in others).

The second phase II study of single agent bortezomib in WM was performed at Dana-Farber Cancer Institute [15], involving 27 patients with relapsed/refractory WM (WMCTG Trial 03-248). The patients received up to eight cycles of bortezomib following the same regimen as for the Canada Clinical Trial. All but one patient had relapsed/refractory disease. Thirteen (48.1%) patients achieved a major response, but no CR was observed. Among major responders, the median time for a 50% reduction in serum IgM was 2.4 months (ranged from 0.9 to 6.1). The median time to progression for all responding patients was 7.9 months (ranged from 3 to 21.4+). Of note, an abrupt IgM spike (>10% increase in serum IgM) was observed for 13 patients immediately following cessation of all bortezomib therapy in this study. For three of four patients who continued to be followed up after showing a spike and not summarily taken off study for progressive disease, the IgM decreased below the previously established nadir with further follow-up.

These studies confirmed that bortezomib treatment was feasible in WM with significant response rates, although there are concerns regarding the incidence of neuropathy.

A third phase II study, coordinated by Dr V. Leblond, of single agent bortezomib should start in France, which will closely monitor neurotoxicity with bortezomib in WM-treated patients. The trial is planned to accrue 34 relapsed/refractory WM patients for 2 cycles of bortezomib at the dose of 1.3 mg/m^2 on days 1, 4, 8, and 11 in a 21-day cycle. Responding patients will continue for up to six cycles, while nonresponding patients will have dexamethasone (20 mg) added at 1, 2, 4, 5, 8, 9, 11, and 12 every 21 days.

3.2 *Bortezomib-Based Regimens in Waldenstrom's Macroglobulinemia*

On the basis of the encouraging data obtained with bortezomib single agent, and in order to improve the response rates, several combination regimens at front line therapy and at relapse/refractory settings are being studied in ongoing trials.

The first combination study was conducted at DFCI [16] in 23 newly diagnosed symptomatic WM patients, and consisted of bortezomib (1.3 mg/m^2, days 1, 4, 8, and 11, every 3 weeks) associated to dexamethasone 40 mg (days 1, 4, 8, and 11) and anti-CD20 monoclonal antibody rituximab (375 mg/m^2) on day 11. Patients received 4 cycles as induction therapy followed by 4 more cycles each given 3 months apart for maintenance therapy. The outcome of this study demonstrated an ORR of 96%, including an 83% major response rate. Importantly, 22% of the patients achieved a CR/nCR and 35% achieved a VGPR or better. These responses compare favorably with those achieved by one of the agents alone (i.e., bortezomib, rituximab, or dexamethasone), for which ORR and major response rates of 40–50% have been reported. Importantly, no CRs/nCRs have been observed with either bortezomib or rituximab alone in the primary treatment setting. An important consideration in this series was the rapid time to response. The median time to a $\geq$25% reduction in serum IgM in this series was 1.4 months, which compares favorably with most of the monotherapies currently in use as well as combination therapies with rituximab with times to response of more than 3–4 months.

Although the median TTP has not been reached in this study, it appears to be more than 30 months compared with that previously reported for either rituximab (14–27 months) or bortezomib (16 months) alone when used as first-line therapy [14, 35–37]. In addition, the response duration with BDR seems to be at least on par with those previously reported with other rituximab-containing regimens, including those in combination with cyclophosphamide, nucleoside analogs, and thalidomide [38, 39].

The most notable toxicity encountered in this study was sensory PN, which was reversible in more than 80% of patients at a median of 6.0 months. The incidence and reversibility of bortezomib-related PN in this study were similar to those previously encountered in single agent experiences with bortezomib in patients with relapsed refractory WM and in other patient populations.

Another phase I/II trial compared bortezomib alone (1.3 mg/m^2, days 1, 4, 8, and 11, every 3 weeks) with bortezomib in combination with the rituximab (375 mg/m^2) with different doses and schedule in patients with relapse/refractory WM [40]. Patients received either bortezomib as above with rituximab administered at day 1 at each cycle for 8 cycles or bortezomib 1.6 mg/m^2, days 1, 8, 15, and 22 of 5-week cycles with rituximab administered at day 1, 8, 15, and 22 of cycles 1 and 4. So far, 17 patients have been treated, and a partial response was observed in 71% of patients, 57% and 80% treated with bortezomib alone and bortezomib in combination with rituximab, respectively. Neurotoxicity was reported in 46% of patients, and the dose of bortezomib was reduced in seven patients and five doses were omitted. The conclusion of the authors is that the weekly schedule is as effective and more convenient than the twice weekly schedule and not more toxic.

Bortezomib has demonstrated efficacy in WM, either at relapse/refractory or in front line settings, and appears to be a new pivotal therapeutic agent in WM. Bortezomib-based regimens will be used in the future in combination; however, the schedule will be of importance to avoid occurrence of neurotoxicity. Ongoing trials focus on determination of the most neuro-protective schedule.

A second phase II trial is conducted at DFCI in 30 evaluable patients with relapsed or refractory WM disease treated to date [41]. All patients received bortezomib IV weekly at 1.6 mg/m^2 on days 1, 8, 15, q 28 days × 6 cycles, and rituximab 375 mg/m^2 at days 1, 8, 15, 22, on cycles 1 and 4 for up to 6 cycles. Complete remission occurred in 1 (3%) and PR occurred in 16 (53%) patients. Most patients achieved response rapidly within 3 months of therapy (2–7 months). Rituximab flare occurred only in six patients (20%). The median duration of response has not been reached, from 3 to 24+ months. Patients tolerated therapy well without significant toxicities, grade-3 peripheral neuropathy occurred in only one patient at cycle 6, and completely resolved within 1 month after stopping the therapy. Attributable toxicities otherwise proved manageable with appropriate supportive care, and the combinations were generally well tolerated. This trial confirms that the combination of weekly bortezomib and rituximab have been well tolerated and demonstrates encouraging activity in relapse/refractory setting. No significant peripheral neuropathy has been observed to date with this regimen. This trial is now open at front line therapy.

Another phase II multicenter study of combination of bortezomib, dexamethasone, and rituximab (BDR) in previously untreated patients with WM is ongoing in Europe (European Myeloma/WM Network; coordinated by Dr. T.M. Dimopoulos). About 61 patients are planned for treatment with BDR administered in one 21-day treatment cycle followed by four 35-day treatment cycles. Bortezomib is administered at a dose of 1.3 mg/m^2 on days 1, 4, 8, and 11 of cycle 1. On cycles 2–5, bortezomib is given at a dose of 1.6 mg/m^2 on days 1, 8, 15, and 22 of each cycle. Only on cycles 2 and 5, following the administration of bortezomib, dexamethasone 40 mg IV and rituximab 375 mg/m^2 IV have to be administered. A total of eight infusions of rituximab will be administered. The trial is ongoing and no data have been released yet.

Finally, 38 newly diagnosed symptomatic WM patients are planned to be enrolled in a phase II study of bortezomib–rituximab followed by cladribine, cyclophosphamide, and rituximab designed to assess response rate and ability to collect stem cells after treatment with bortezomib and rituximab (M.D. Anderson Cancer Center, coordinated by Dr. S. Thomas). The schedule proposed is bortezomib (1.6 mg/m^2 IV weekly on days 1, 8, 15, and 22) and rituximab (375 mg/m^2 IV on day 8 and 22) on 35-day cycles for 2 cycles, followed for responders by 1 cycle of cladribine given IV once a day on days 1–5 with cyclophosphamide taken orally, twice a day on days 1–5, and rituximab given IV once a week for 4 weeks.

4 Conclusion

In summary, the last decade has marked a new era in the treatment of monoclonal gammophaties. Indeed, a new paradigm shift has evolved utilizing novel therapeutic agents, such as bortezomib, targeting the malignant clone and the surrounding bone marrow microenvironment. The combination of novel agents with chemotherapeutic drugs and/or glucocorticoids has demonstrated high response rates with complete remission rates comparable with those achieved in the stem cell transplant setting. This has been supported by in vitro and in vivo evidences showing the antitumor activity of these novel agents in WM, as well as in multiple myeloma and other B cell malignancies. The future holds many more challenges for the treatment of MM and WM. These include combination of agents that achieve higher responses and longer survival, individualized therapies that are based on genetic and molecular abnormalities present in each patient, and clinical trials to test the benefit of novel agents in comparison and in addition to autologous stem cell transplantation, as well as other conventional approaches. Together, these therapies should lead to higher response rates, more durable duration of response, less toxicity, and prolonged survival for patients, making plasma cell discrasia an increasingly chronic and treatable disease.

References

1. Owen RG, Treon SP, Al-Katib A, Fonseca R, Greipp PR, McMaster ML et al (2003) Clinicopathological definition of Waldenstrom's macroglobulinemia: consensus panel recommendations from the Second International Workshop on Waldenstrom's macroglobulinemia. Semin Oncol 30:110–115
2. Herrinton LJ, Weiss NS (1993) Incidence of Waldenstrom's macroglobulinemia. Blood 82:3148–3150
3. Jemal A, Murray T, Ward E, Samuels A, Tiwari RC, Ghafoor A et al (2005) Cancer statistics, 2005. CA Cancer J Clin 55(1):10–30
4. Ghobrial IM, Gertz MA, Fonseca R (2003) Waldenstrom macroglobulinaemia. Lancet Oncol 4:679–685
5. Ghobrial IM, Fonseca R, Gertz MA, Plevak MF, Larson DR, Therneau TM et al (2006) Prognostic model for disease-specific and overall mortality in newly diagnosed symptomatic patients with Waldenstrom macroglobulinaemia. Br J Haematol 133:158–164
6. Benjamin M, Reddy S, Brawley OW (2003) Myeloma and race: a review of the literature. Cancer Metastasis Rev 22(1):87–93
7. McMaster ML (2003) Familial Waldenstrom's macroglobulinemia. Semin Oncol 30:146–152
8. Kyle RA, Therneau TM, Rajkumar SV, Offord JR, Larson DR, Plevak MF et al (2003) Long-term follow-up of IgM monoclonal gammopathy of undetermined significance. Semin Oncol 30:169–171
9. Morra E, Cesana C, Klersy C, Barbarano L, Varettoni M, Cavanna L et al (2004) Clinical characteristics and factors predicting evolution of asymptomatic IgM monoclonal gammopathies and IgM-related disorders. Leukemia 18:1512–1517
10. Kyle RA, Treon SP, Alexanian R, Barlogie B, Bjorkholm M, Dhodapkar M et al (2003) Prognostic markers and criteria to initiate therapy in Waldenstrom's macroglobulinemia: consensus panel recommendations from the Second International Workshop on Waldenstrom's macroglobulinemia. Semin Oncol 30:116–120
11. Kriangkum J, Taylor BJ, Treon SP, Mant MJ, Belch AR, Pilarski LM (2004) Clonotypic IgM V/D/J sequence analysis in Waldenstrom macroglobulinemia suggests an unusual B-cell origin and an expansion of polyclonal B cells in peripheral blood. Blood 104:2134–2142
12. Schop RF, Van Wier SA, Xu R, Ghobrial I, Ahmann GJ, Greipp PR et al (2006) 6q deletion discriminates Waldenstrom macroglobulinemia from IgM monoclonal gammopathy of undetermined significance. Cancer Genet Cytogenet 169:150–153

13. Dimopoulos MA, Anagnostopoulos A, Kyrtsonis MC, Castritis E, Bitsaktsis A, Pangalis GA (2005) Treatment of relapsed or refractory Waldenstro¨ms macroglobulinemia with bortezomib. Haematologica 90:1655–1658
14. Chen CI, Kouroukis CT, White D, Voralia M, Stadtmauer E, Stewart AK et al (2007) Bortezomib is active in patients with untreated or relapsed Waldenstro¨ms macroglobulinemia: a phase II study of the National Cancer Institute of Canada Clinical Trials Group. J Clin Oncol 25:1570–1575
15. Treon SP, Hunter ZR, Matous J, Joyce RM, Mannion B, Advani R et al (2007) Multicenter clinical trial of bortezomib in relapsed/refractory Waldenstro¨ms macroglobulinemia: results of WMCTG trial 03-248. Clin Cancer Res 13:3320–3325
16. Treon SP, Ioakimidis L, Soumerai JD et al (2009) Primary therapy of Waldenstrom macroglobulinemia with bortezomib, dexamethasone, and rituximab: WMCTG clinical trial 05-180. J Clin Oncol 27:3830–3835
17. Williams S, Pettaway C, Song R et al (2003) Differential effects of the proteasome inhibitor bortezomib on apoptosis and angiogenesis in human prostate tumor xenografts. Mol Cancer Ther 2:835
18. Albanell J, Baselga J, Guix M et al (2003) Phase I study of bortezomib in combination with docetaxel in anthracycline-pretreated advanced breast cancer. Proc Am Soc Clin Oncol 22:16
19. Hideshima T, Chauhan D, Richardson P et al (2002) NF-kappa B as a therapeutic target in multiple myeloma. J Biol Chem 277:16639
20. Gilmore TD, Koedood M, Piffat KA et al (1996) Rel/NF-kappaB/IkappaB proteins and cancer. Oncogene 13:1367
21. Mitsiades N, Mitsiades CS, Poulaki V et al (2002) Biologic sequelae of nuclear factor-kappaB blockade in multiple myeloma: therapeutic applications. Blood 99:4079
22. Hideshima T, Chauhan D, Schlossman R et al (2001) The role of tumor necrosis factor alpha in the pathophysiology of human multiple myeloma: therapeutic applications. Oncogene 20:4519
23. Mitsiades N, Mitsiades CS, Poulaki V et al (2002) Molecular sequelae of proteasome inhibition in human multiple myeloma cells. Proc Natl Acad Sci USA 99:14374
24. Hideshima T, Richardson P, Chauhan D et al (2001) The proteaosme inhibitor PS-341 inhibits growth, induces apoptosis, and overcomes drug resistance in human multiple myeloma. Cancer Res 61:3071
25. Shah SA, Potter MW, McDade TP et al (2001) 26S proteasome inhibition induces apoptosis and limits growth of human pancreatic cancer. J Cell Biochem 82:110
26. Ling YH, Liebes L, Mg B et al (2002) PS-341, a novel proteasome inhibitor, induces Bcl-2 phosphorilation and cleavage in association with GM-2 phase arrest and apoptosis. Mol Cancer Ther 1:841
27. Hideshima T, Mitsiades C, Akiyama M et al (2003) Molecular mechanisms mediating antimyeloma activity of proteasome inhibitor PS-341. Blood 101:1530
28. Roccaro A, Leleu X, Sacco A et al (2008) Dual targeting of the proteasome regulates survival and homing in Waldenstrom macroglobulinemia. Blood 111:4752–4763
29. Mitsiades CS, Treon SP, Mitsiades N et al (2002) TRAIL/Apo2L ligand selectively induces apoptosis and overcomes drug resistance in multiple myeloma: therapeutic implications. Blood 99:4525
30. Drexler HC, Risau W, Konerding MA (2000) Inhibition of proteasome function induces programmed cell death in proliferating endothelial cells. FASEB J 14:65
31. Oikawa T, Sasaki T, Nakamura M et al (1998) The proteasome is involved in angiogenesis. Biochem Biophys Res Commun 246:243
32. Sunwoo JB, Chen Z, Dong G et al (2001) Novel proteasome inhibitor PS-341 inhibits activation of nuclear factor-kappa B, cell survival, tumor growth, and angiogenesis in squamous cell carcinoma. Clin Cancer Res 7:1419
33. LeBlanc R, Catley LP, Hideshima T et al (2002) Proteasome inhibitor PS-341 inhibits human myeloma cell growth in vivo and prolongs survival in a murine model. Cancer Res 62:4996

34. Roccaro AM, Hideshima T, Raje N et al (2006) Bortezomib targets multiple myeloma endothelial cells. Cancer Res 1:184
35. Gertz MA, Rue M, Blood E et al (2004) Multicenter phase 2 trial of rituximab for Waldenstrom macroglobulinemia (WM): an Eastern Cooperative Oncology Group Study (E3A98). Leuk Lymphoma 45:2047–2055
36. Dimopoulos MA, Zervas C, Zomas A et al (2002) Treatment of Waldenstrom's macroglobulinemia with rituximab. J Clin Oncol 20:2327–2333
37. Treon SP, Emmanouilides C, Kimby E et al (2005) Extended rituximab therapy in Waldenstro¨ m's macroglobulinemia. Ann Oncol 16:132–138
38. Treon SP, Soumerai JD, Branagan AR et al (2008) Thalidomide and rituximab in Waldenstrom's macroglobulinemia. Blood 112:4452–4457
39. Buske C, Hoster E, Dreyling M et al (2009) The addition of rituximab to front-line therapy with CHOP (R-CHOP) results in a higher response rate and longer time to treatment failure in patients with lymphoplasmacytic lymphoma: results of a randomized trial of the German Low-Grade Lymphoma Study Group (GLSG). Leukemia 23:153–161
40. Anagnostopoulos A, Dimopoulos MA, Aleman A, Weber D, Alexanian R, Champlin R et al (2001) High-dose chemotherapy followed by stem cell transplantation in patients with resistant Waldenstrom's macroglobulinemia. Bone Marrow Transplant 27(10):1027–1029
41. Tournilhac O, Leblond V, Tabrizi R, Gressin R, Senecal D, Milpied N et al (2003) Transplantation in Waldenstrom's macroglobulinemia – the French experience. Semin Oncol 30 (2):291–296

Bortezomib in Systemic Light-Chain Amyloidosis

Morie A. Gertz and Raymond Comenzo

Abstract The treatment for immunoglobin light-chain amyloidosis has included high-dose chemotherapy with stem cell transplantation. This effective technique is only applicable to approximately 20% of patients. Standard dose chemotherapy has included melphalan with dexamethasone. This regimen results in response rates in excess of 50%. With the introduction of IMID therapy, the combination of thalidomide or lenalidomide with dexamethasone has produced respectable response rates, but tolerance of patients with amyloidosis to IMID's is less than in patients with multiple myeloma. The introduction of bortezomib for the management of light-chain amyloidosis is reviewed in this chapter. Bortezomib is highly active both as a single agent and in combination with corticosteroids in the treatment of immunoglobulin light-chain amyloidosis, with high hematologic and organ response rates reported. The agent can be used with renal insufficiency, including dialysis dependence. The agent has not been extensively tested in patients with class 3–4 heart failure.

1 Introduction

The systemic amyloidoses are a group of fatal diseases in which the misfolding of extracellular protein has a prominent role in pathology [1]. Amyloid fibrils are derived from an assembly of specific precursor proteins, the identity of which determines the type of amyloid that a patient has. All amyloid fibrils deposit with a β-sheet secondary structure and are linear and non-branching, 7–10 μm in

M.A. Gertz (✉)
Division of Hematology, Mayo Clinic, Rochester, MN 55905, USA
e-mail: gertz.morie@mayo.edu

R. Comenzo
Hematology Service, Memorial Sloan Kettering Cancer Center, New York, NY 10021, USA
e-mail: comenzor@mskcc.org; RComenzo@tuftsmedicalcenter.org

I.M. Ghobrial et al. (eds.), *Bortezomib in the Treatment of Multiple Myeloma*, Milestones in Drug Therapy,
DOI 10.1007/978-3-7643-8948-2_10,

diameter, with a distinctive appearance by electron microscopy (EM). In tissue biopsies stained with Congo red and viewed microscopically under polarized light, amyloid has a pathognomonic red-to-green dichroism. In contrast, the deposits of immunoglobulin light-chain deposition disease by EM are not fibrillar but diffuse and granular and do not exhibit birefringence in polarized light [2]. In systemic light-chain amyloidosis (AL), the precursor protein is usually an abnormal monoclonal immunoglobulin light chain that is the product of a clone of plasma cells. Five to ten percent of patients with multiple myeloma, for example, have coincident AL as do patients with Waldenstrom's macroglobulinemia [3]. The majority of patients with AL, however, have "dangerous small" clones and do not meet criteria for myeloma [4, 5].

The introduction of bortezomib as a novel agent for the management of AL has shown promising results over the past 2 years. However, no phase III studies exist on its use in AL. Therefore, the current literature necessitates a comparison of phase I and II studies that suffer from issues of patient selection and relative proportions of patients with cardiac, renal, hepatic, and peripheral nervous system amyloid. To assist in understanding the current (and future) literature on the use of bortezomib in AL, we begin with a review of the prognostic features associated with AL and criteria for hematologic and organ responses. It is important to note that interpreting the response to treatment is different in AL than in multiple myeloma. In the latter, the total tumor mass determines the outcome. Since the monoclonal immunoglobulin is a surrogate for the number of plasma cells, responses can simply be interpreted by the reduction in the M protein [6, 7]. In AL, relevant baseline prognostic factors are related to the production and concentration of amyloid-forming light chains and the extent of organ dysfunction, and response criteria cover both the monoclonal protein, usually measured as a free light chain, not an electrophoretic M protein, as well as objective improvements in organ function [8]. A review of the established prognostic and response criteria and of the history of therapy for AL is warranted before discussing the current data on bortezomib for AL.

2 Prognostic Features of AL

When a patient presents with a syndrome compatible with amyloidosis, is found to have a monoclonal protein consistent with AL, and has histologic confirmation of the diagnosis, assessment of prognosis represents the next step in management [9]. Virtually all patients with amyloidosis succumb to complications related to cardiac involvement. This can be either progressive cardiomyopathy with intractable heart failure or sudden death due to ventricular fibrillation, ventricular asystole, or unrecognized pulmonary emboli originating from the left atrial appendage [10, 11]. The most important determinant of clinical outcome is the extent of myocardial involvement.

Congestive heart failure diagnosed clinically is associated with a median survival in patients with AL of only 6 months [12]. Exertional syncope is a predictor of early death [13, 14]. Echocardiography remains an important tool for assessing the prognosis of patients with AL. Two-dimensional Doppler echocardiography shows that 40% of AL patients have cardiac involvement, although only 17% have symptoms. In a study of 64 patients, cardiac relaxation was abnormal in early-stage AL but, when advanced, restrictive filling and a shortened deceleration time developed [15, 16]. In another study, 40 AL patients who had symptomatic cardiac involvement at diagnosis, and a median brain natriuretic peptide (BNP) of 1,000 ng/ml (range 100–4,510), were treated uniformly with oral therapy [17]. Their overall survival was influenced by baseline interventricular septal thickness (IVST) as measured by echocardiogram. Those with IVST $\geq$ or <0.14 cm lived medians of 10 and 22 months respectively ($P = 0.02$).

The cardiac biomarkers troponin and N-terminal pro-brain natriuretic peptide (NT-proBNP) are objective measures of cardiac status, unlike echocardiography that potentially has operator bias, and have demonstrated important predictive value for survival in AL [18–20]. With the use of these biomarkers, thresholds that correlate with distinct survivals have been established for troponin T, troponin I and NT-proBNP, allowing the designation of three stages of cardiac amyloid (I, II and III). The thresholds for troponin T, troponin I and NT-proBNP are <0.035 μg/L, <0.1 μg/L, and <332 ng/L, respectively. In patients with stage I disease neither the troponin (using either T or I) nor the NT-proBNP value exceeds threshold, while with stage II one of the values, and with stage III both of them, exceed the thresholds. Median survivals decrease with increasing stage, from 26 to 11 to 4 months respectively. The availability of the cardiac biomarkers and this staging system allow objective assessment of cardiac amyloid for both patient-care and clinical research purposes. At Mayo Clinic, for example, we have recently introduced the serum troponin T level as a criterion for stem cell transplantation(SCT) [21]. If the troponin T level is ≥ 0.06 μg/mL, we now exclude these patients from transplantation because the 100-day all-cause mortality in patients with troponin T ≥ 0.06 μg/mL was 28% as compared to 7% in those with levels <0.06 μg/mL ($P < 0.001$). Cardiac biomarker staging is also important for the design and conduct of clinical trials in order to facilitate more consistent and reliable comparisons of therapeutic outcomes.

In a uniformly treated cohort of 153 patients with AL reported in 1989, the median survival of the entire group was 20 months; 5-year survival was 20% [22]. Patients with congestive heart failure lived a median of 8 months and only 2.4% of them were alive at 5 years. In contrast, patients with only peripheral neuropathy at diagnosis lived a median of 40 months and 32% of them were alive at 5 years. Classification of patients with AL into four groups based on dominant organ involvement at presentation (cardiac, renal, liver/GI, peripheral nervous system) remains useful. Women have a slightly longer survival than men even in the setting of symptomatic cardiac involvement [23].

Referral bias exists at large treatment centers for amyloidosis, and differences exist between centers. In order to be treated at a center of excellence, the patient must be physically able to travel. This needs to be considered when interpreting the results of clinical trials reported from a single center. In AL patients evaluated at the

Mayo Clinic, the median survival was 2 years. If the cohort was limited to patients who were evaluated within 30 days of diagnosis, however, survival decreased to only 13 months [24]. When two cohorts of AL patients were compared between the center at Mayo Clinic and that at The Amyloidosis Center in Pavia, Italy, the median survival of the Italian patients was 30 months vs. 12 months in the Mayo cohort. Renal involvement was nearly twice as common in the Italian cohort, and there was a significantly higher prevalence of cardiac amyloidosis at Mayo. Such differences must be taken into account when comparing single-center reports of therapy [25].

3 Assessing Response in AL

Oftentimes, assessing the response to therapy in AL can be difficult. Unlike multiple myeloma in which a quantifiable M protein can be used as a surrogate for reduction in tumor mass, interpreting the response of patients with AL is more complex because the majority does not have a measurable M protein and the clonal light chain and not the classical M protein provides the precursor protein. Only 10% of AL patients have a serum monoclonal protein >1.5 g/dL. Many of the patients have only light chain proteinemia; quantification has been challenging until the advent of the quantitative serum free light chain assay. Of note, quantification of the urinary light chain is difficult because a high proportion of patients have marked albuminuria that overwhelms the monoclonal urine protein peak rendering it unquantifiable and because many also have renal tubular injury that causes spillage of non-pathologic light chains and Ig fragments making the late fraction protein peak insensitive for tracking response. In addition, patients with AL have an average of only 5% plasma cells in the bone marrow before therapy. Although this degree of plasmacytosis is recognized to be abnormal, it is difficult in serial biopsies to consistently determine if there has been a change in the actual percentage of plasma cells in the bone marrow without reliable sampling and reproducible immunohistochemical identification of light-chain restricted plasma cells.

Consensus criteria have been developed for assessing response in AL but these criteria thus far have failed to incorporate new information on cardiac biomarkers. Despite these limitations, a hematologic response to treatment in AL has been defined as a 50% reduction in the precursor monoclonal protein level [8]. The effect of a 50% decrease in the monoclonal protein level in a cohort of patients receiving high-dose therapy followed by stem cell transplantation is associated with improved survival [26].

The nephelometric serum immunoglobulin free light chain assay has become increasingly important in quantifying hematologic responses in patients with amyloidosis. The light chain assay is both a sensitive and specific marker for AL [27]. The absolute value of the pathologic free light chain has prognostic value. Patients with higher baseline levels of pathologic free light chains have a higher risk of death (hazard ratio of 2.6), and achievement of a free light chain response is a better predictor of survival than a standard defined complete hematologic response [28]. Reducing the level of pathologic free light chains by greater than 50% has been associated with substantial survival benefit regardless of the type of chemotherapy used. Marked clinical and organ improvement may be delayed, but the treatment strategy can be guided by the early effect of therapy on the free light chain level [27]. It has been shown that survival is better in those patients who have a 50% reduction compared with those who have <50% reduction in free light chain values. Patients who have normalization have the best outcomes [29].

The strategy behind therapy and the most important clinical endpoint in AL is functional improvement and organ response. The underlying hypothesis is that amyloid deposits in organs cannot resolve until production of the precursor light chain has been interrupted. After precursor protein production is disrupted, the amyloid deposits, as demonstrated by SAP component scintigraphy, can resolve if the organ has not been irrevocably damaged by the fibrillar burden [30]. There is a prevalent hypothesis suggesting that symptoms and organ dysfunction in AL are in part consequences of soluble intermediate forms of the misfolded precursor protein (so-called "toxic intermediates") rather than the insoluble amyloid fibril [31]. If this is indeed the case, interruption in precursor protein production can lead to a reduction in toxic intermediates and improvement in organ function without regression of the amyloid deposits.

There is an excellent correlation between a hematologic response to therapy and subsequent organ response and survival [27]. Therefore, a 50% reduction in the serum or urine monoclonal protein or the nephelometric serum free light chain assay is an important surrogate for organ response and is expected to lead to functional organ improvement if the organ has not been severely damaged. It remains to be determined whether the final endpoint of therapy should be a 50% reduction or complete abrogation of pathologic light chain production. Outcome may partially depend on the intrinsic amyloid-forming propensity of the light chain. Light chains that have only a moderate tendency to form insoluble fibrils may require only a modest reduction in concentration to shift the equilibrium constant from misfolded intermediates back to soluble precursor protein. Light chains that misfold easily into an amyloid-forming conformation likely require near complete elimination to prevent amyloid deposition and organ disease progression. Unfortunately, there is currently no way to measure the toxicity or amyloid-forming propensity of an individual patient's pathologic light chains.

The definition of an organ response is a 50% reduction in urinary albumin loss in patients with renal amyloidosis, a decrease in liver size or a 50% reduction in serum alkaline phosphatase level in patients with hepatic amyloidosis, and echocardiographic improvement in patients with cardiac amyloidosis, although recently reductions in the BNP level by 30% have been proposed as evidence of a cardiac response. The immunoglobulin free light chain level is monitored serially in all patients receiving therapy. It is conventionally accepted that a 50% reduction constitutes a partial response. Normalization of the free light chain level and negative immunofixation of the serum and urine is now considered a complete hematologic response.

4 Therapy of AL

In the absence of therapy that is capable of solubilizing the amyloid fibril or disrupting the dynamic equilibrium between soluble light chains and toxic intermediates, all the available therapies for AL have been directed against interruption of the synthesis of the precursor protein, the light chain. Therefore, all therapies have been directed at reducing the clonal plasma cells. Melphalan-based stem cell transplant, melphalan and dexamethasone, and thalidomide- and lenalidomide-based therapies have all been used with varying success in the treatment of AL.

4.1 High Dose Chemotherapy with Autologous Hematopoietic Stem Cell Transplantation

The effectiveness of SCT in reversing the clinical manifestations of AL in most surviving patients has been documented at numerous centers. The toxicities of SCT have also been documented. In one series of 394 patients eligible for SCT, 63 declined SCT and 19 had progressive disease before stem cell mobilization with G-CSF alone, making them ineligible for SCT [28]. Stem-cell mobilization was performed in 312 patients but 35 of them could not continue with SCT because of death or complications of mobilization. Of the 277 who underwent stem cell transplant, 36 (13%) died of treatment-related causes. However, for the 312 patients who underwent stem cell mobilization in this series, the median overall survival post-SCT was 4.6 years with 47% surviving at 5 years. For patients without cardiac involvement at SCT the median survival was 6.4 years with 60% 5-year survival, while for those with cardiac involvement at SCT the median survival was 1.6 years with 29% 5-year survival. Forty-four percent of evaluable patients achieved organ responses at 1-year post-SCT. The complete response rate was 40% and was associated with an 82% 5-year survival compared to the 55% for those not achieving complete response.

In another series of 270 patients who underwent SCT, the overall treatment-related mortality was 11% [32]. In a proportional hazards model, predictors of outcome included number of organ systems affected and the baseline FLC level. Patients achieving a complete or partial response had significantly better survival than those with no response (median survival <12 months).

4.2 Melphalan and Dexamethasone

In 2003, investigators in Pavia, Italy, reported on their experience in a phase II trial using oral melphalan and dexamethasone (MDex) in patients not eligible for stem cell transplant ($n = 46$) [33]. They demonstrated a hematologic response rate of 67% with 33% complete hematologic responses and an organ response rate of 48%. There were two treatment-related deaths in the first 100 days of therapy and two patients subsequently developed myelodysplasia. Thirty-two of the 46 patients had advanced cardiac involvement and 8 of the 32 died within 5 months of starting therapy. An updated report that followed surviving patients for a median of 5 years showed median progression-free and overall survival rates of 3.8 and 5.1 years respectively.

Recently, we reported a series of 40 consecutive patients with advanced cardiac amyloid treated with the same regimen [17]. Nine of the 40 patients died within 3 months of starting therapy as the result of sudden cardiac death ($n = 6$) or infection ($n = 3$). The overall response rate was 58% with 13% complete hematologic responses. Median overall survival was 10.5 months and baseline predictors of survival included gender, troponin I level and interventricular septal thickness. The most significant predictor of survival was response to therapy. Responders lived a median of 22 and non-responders 9 months.

4.3 MDex vs. SCT

Investigators in the Myélome Autogreffe and Intergroupe Francophone du Myélome Intergroup conducted a randomized prospective multi-center phase III trial ($n = 100$) in which SCT with standard iv melphalan was compared with oral MDex given continuously for up to 18 months [34]. In the SCT group, there was a 44% early failure rate largely due to treatment-related deaths and progression of disease. Analysis of outcomes between the groups demonstrated a center effect in that patients treated at the majority of centers (27 of 29) had far better survival if they received MDex ($p \ll 0.01$) while patients in the two centers in which the largest fraction of SCT were performed had a trend of better survival with SCT. Comparisons of response rates and survival between those alive at least 3 months post-SCT and those who completed at least 3 months of MDex showed no difference. For both groups the hematologic response rates were 65%. Median survival was 48 months for SCT and 58 months for MDex. This phase III trial did not clearly define a standard therapy for AL but made the point that sicker AL patients do not benefit from the SCT approach even when dose-adjusted melphalan is used.

4.4 Thalidomide and Lenalidomide

In the treatment of plasma cell dyscrasias we have entered an era of novel therapies, calling into question the continued need for SCT in both MM and AL, and creating increased awareness of the risk of myelodysplasia in AL patients receiving oral melphalan [35, 36]. Phase III clinical trials in multiple myeloma comparing standard MP to thalidomide or bortezomib combined with MP have shown increased activity and improved survival for those receiving combination therapies including the novel agents [37, 38]. Therefore, in AL, the study of MDex combined with novel agents is clearly warranted and may lead to comparisons with SCT-based therapies in the future.

Both thalidomide and lenalidomide are immunomodulatory oral agents that inhibit TNF-α and have multiple affects on immune cells and the bone marrow microenvironment [39]. The mechanism of action in myeloma and AL is not definitively known. Overall lenalidomide is better tolerated than thalidomide because it does not cause as much sedation and neuropathy, and because it is administered only 3 weeks a month, allowing a 1-week holiday. Both agents carry risks of constipation, fluid retention and thromboembolic events. Single agent thalidomide has been shown to be toxic and ineffective, but used as a salvage regimen with dexamethasone, activity has been demonstrated with hematologic response rates of 48% with 19% complete responses; the response rate was higher in those taking higher doses of thalidomide [40]. Median time to response was 3.6 months. Organ response rate was 26%. Fluid retention and symptomatic bradycardia without QT prolongation were common adverse reactions while neuropathic and thromboembolic complications were rare.

In a phase II trial that tested the combined approach of SCT and adjuvant thalidomide and dexamethasone after SCT, nearly half of the patients receiving adjuvant dexamethasone +/−thalidomide had improved responses at 12 months including six complete responses [41]. The response rate at 12 months was 78% with 39% complete responses. Baseline characteristics associated with increased overall survival were the number of organ systems involved and the troponin I. A normal serum free light chain κ-to-λ ratio at 3 months post-HCT and achievement of a hematologic response at 12 months post-HCT were associated with improved overall and progression-free survival. Further study of such combined approaches continues with SCT as a platform for therapy.

Two phase II trials have been conducted with lenalidomide +/− dexamethasone in AL [42, 43]. In one, patients received lenalidomide for 3 months, and if they did not respond, dexamethasone was added [44]. Of the 23 patients enrolled, over half had been treated in the past. Within the first 3 cycles, ten patients stopped treatment, including four who had early deaths. But ten patients responded to treatment, all but one of whom had dexamethasone added to the regimen. Grade 3/4 adverse events possibly attributable to lenalidomide were neutropenia (45%), thrombocytopenia (27%), rash (18%), and fatigue (18%). In a second phase II trial, 34 patients with AL were treated, almost all of whom had prior therapy (19 had had one or two SCT) [45]. Sixteen of 24 evaluable patients achieved a hematologic response, including seven who achieved a complete hematologic response. Over a third of patients experienced grade 3/4 fatigue and myelosuppression and 10% experienced thromboembolic complications. There were no thromboembolic complications after aspirin prophylaxis was added.

5 Bortezomib for AL

Bortezomib is a first-in-class proteasome inhibitor. By inhibiting proteasome function in MM cells, bortezomib causes a "traffic-jam in the ER" and thereby activates the stress-activated protein kinase (SAPK)/JNK pathway and JNK-mediated mitochondrial apoptotic signaling [46]. Some have conjectured that the clonal plasma cells in AL are subject to intracellular stress because amyloid-forming light chains can produce a load for the endoplasmic reticulum that makes them more sensitive to proteasome inhibition [47]. The rationale for pursuing bortezomib in AL is that the drug induces a rapid high response rate and that the extent and rapidity of monoclonal light chain reduction closely relates to outcome. Moreover, bortezomib can induce rapid renal improvement targeting NF-κ-B activation, may interrupt the process of tubulointerstitial injury and result in improved kidney function [48, 49].

The first reports on the use of bortezomib in the management of amyloidosis did not appear until the American Society of Hematology Meeting held in 2006. At that time, a case report described a patient with hepatic AL previously treated with melphalan, dexamethasone, and thalidomide who received bortezomib with a reduction in hepatic size from 20 to 10 cm indicative of an organ response with disappearance of the monoclonal protein (hematologic response) and improved cardiac and renal function. Also, the National Amyloidosis Center in Great Britain reported preliminary information on 18 patients treated with bortezomib, half of whom also received dexamethasone, and all of whom had received prior thalidomide-based therapy. A hematologic response was seen in 77% of the patients, 16% complete, with 27% organ responses. This experience [50] was subsequently published in 2008, updated to 20 patients with a median of three lines of prior chemotherapy, describing a response rate to bortezomib of 80% with 15% complete responses. Forty percent of patients had to discontinue therapy due to treatment-related toxicity [42, 50, 51].

At the American Society of Clinical Oncology meeting in 2007, Reece and colleagues presented initial results of the phase I component of an international phase I/II study of single-agent bortezomib, with results up-dated at a meeting later that year [43]. The phase I component included seven cohorts covering different doses on two schedules, a weekly schedule (days 1, 8, 15 and 22 on a 35-day cycle) and a bi-weekly schedule (the standard 21-day cycle). On the weekly schedule, dose-limiting toxicity was not seen even at 1.6 mg/m^2, now considered the maximum tolerated dose for the weekly schedule. This dose exceeds the standard recommendation for multiple myeloma. For bi-weekly, the maximum tolerated dose was found to be 1.3 mg/m^2. These doses have been approved by an international data monitoring committee and are being used in the phase II component of the study [43].

In the phase I component, gastrointestinal side effects such as ileus, constipation, nausea, vomiting and diarrhea were prominent. Grades 3 and 4 adverse events were seen in 42% of patients, fatigue in 17%, infection in 17%, heart failure in 8%, and retinal detachment in 8%. Since this was a phase I study with dose escalation, there were no responses in cohort 1 (lowest dose of bortezomib) and only one partial response in cohort 2. Of the 13 patients in cohorts 3–6, there were eight responses (62%). Organ responses were actually seen in three cardiac, five renal and two peripheral nervous system patients despite the recognized neurotoxicity associated with bortezomib therapy. The conclusion was that bortezomib had significant activity in relapsed AL with good tolerance.

In a recently reported study, 18 patients including seven who had relapsed or progressed after prior therapy were treated with bortezomib and dexamethasone [52]. Sixteen of the 18 patients were evaluable. Eleven patients (61%) had two or more organs involved. Renal and cardiac involvement was seen in 14 and 15 patients respectively. Serum creatinine was elevated in six. The bortezomib schedule was 1.3 mg/m^2 on days 1, 4, 8, and 11 with dexamethasone 40 mg for four days every 21, and a median of five cycles were administered. A hematologic response was seen in 15 patients (94%), and a hematologic complete response in 7 (44%), including all five patients who had failed prior high-dose dexamethasone. One quarter of patients had a response in at least one affected organ. Responses were achieved at a median of 0.9 months, and median time to organ response was 4 months. The toxicities associated with bortezomib combined with dexamethasone were not unique and included neurotoxicity, fatigue, peripheral edema, constipation, and exacerbation of postural hypotension. Dose modification or cessation was required in 11 patients. The toxicities were manageable with close monitoring and dosage adjustments.

There are three critical areas that require further study. First, it is important to emphasize that the rapidity of response creates a dilemma with respect to how long treatment should be continued beyond maximal response. Of course, the rapidity of hematologic response may be important in the management of patients with AL since this may disrupt the progressive loss of organ function. However, our understanding of the significance of the rapid response is limited at this time. Does it reflect a cytotoxic effect of the drug or is it some type of cell-suppressive effect? Second, we do not know how durable responses will be and the degree to which prolonged treatment with bortezomib beyond maximal response will enhance response durability – the other half of the dilemma. Third, we do not know if there are subsets of AL patients who will not be able to tolerate the drug. In the soon-to-be completed phase I/II trial discussed above, for example, patients who experience even a short run of non-sustained ventricular tachycardia on 24-h Holter study are excluded from therapy. Therefore, the tolerability of bortezomib in patients with different stages of cardiac amyloid remains to be determined. In contrast, bortezomib is known to be safe in patients with renal dysfunction [53]. Renal involvement is very common in amyloidosis, so a non-nephrotoxic regimen is useful. Nevertheless, on the phaseI/II, trial patients needed a creatinine clearance of $\geq$40 ml/min to be treated.

Because of the relatively recent introduction of bortezomib to treat amyloidosis, the follow-up of most patients is less than 1 year. It is also possible, since organ responses are time dependent, that as more time elapses, the actual organ response rates will increase [54]. In myeloma, Bortezomib has been added to high-dose melphalan conditioning, and since SCT is highly effective in the management of AL, the two may act synergistically to improve response rate. Similarly, in non-transplant myeloma patients, bortezomib has been tested with oral melphalan and prednisone (BMP) against MP alone, with BMP showing significant benefits in overall and progression-free survival. In AL, the addition of bortezomib to oral MDex should be evaluated in the near future, hopefully in a phase III trial vs. standard MDex.

6 Conclusion

The limited data on the use of bortezomib in the treatment of amyloidosis are highly encouraging with hematologic and organ responses being achieved rapidly. The agent may be used in patients with all levels of renal failure including dialysis dependence but should be used cautiously in patients with advanced cardiac involvement. Because the various prognostic features vary from series to series and the criteria for response are evolving over time, caution in interpreting results is warranted and clinical research studies will need to continue in order to define the role of proteasome inhibition in the management of AL particularly with respect to combining bortezomib with standard agents such as melphalan and the optimal duration of bortezomib therapy beyond maximal hematologic response.

References

1. Merlini G, Westermark P (2004) The systemic amyloidoses: clearer understanding of the molecular mechanisms offers hope for more effective therapies. J Intern Med 255:159–178
2. Salant DJ, Sanchorawala V, D'Agati VD (2007) A case of atypical light chain deposition disease–diagnosis and treatment. Clin J Am Soc Nephrol 2:858–867
3. Gertz MA, Merlini G, Treon SP (2004) Amyloidosis and Waldenstrom's macroglobulinemia. Hematology Am Soc Hematol Educ Program 257–282.
4. Merlini G, Stone MJ (2006) Dangerous small B-cell clones. Blood 108:2520–2530
5. Comenzo RL (2007) Managing systemic light-chain amyloidosis. J Natl Compr Canc Netw 5:179–187
6. Blade J, Samson D, Reece D et al (1998) Criteria for evaluating disease response and progression in patients with multiple myeloma treated by high-dose therapy and haemopoietic stem cell transplantation. Myeloma Subcommittee of the EBMT. European Group for Blood and Marrow Transplant. Br J Haematol 102:1115–1123
7. Durie BG, Harousseau JL, Miguel JS et al (2006) International uniform response criteria for multiple myeloma. Leukemia 20:1467–1473
8. Gertz MA, Comenzo R, Falk RH et al (2005) Definition of organ involvement and treatment response in immunoglobulin light chain amyloidosis (AL): a consensus opinion from the 10th International Symposium on Amyloid and Amyloidosis, Tours, France, 18–22 April 2004. Am J Hematol 79:319–328
9. Gertz MA (2004) The classification and typing of amyloid deposits. Am J Clin Pathol 121:787–789

10. Palladini G, Malamani G, Co F et al (2001) Holter monitoring in AL amyloidosis: prognostic implications. Pacing Clin Electrophysiol 24:1228–1233
11. Feng D, Edwards WD, Oh JK et al (2007) Intracardiac thrombosis and embolism in patients with cardiac amyloidosis. Circulation 116:2420–2426
12. Sawyer DB, Skinner M (2006) Cardiac amyloidosis: shifting our impressions to hopeful. Curr Heart Fail Rep 3:64–71
13. Kawakami K, Abe H, Harayama N, Nakashima Y (2003) Successful treatment of severe orthostatic hypotension with erythropoietin. Pacing Clin Electrophysiol 26:105–107
14. Vita G, Mazzeo A, Di Leo R, Ferlini A (2005) Recurrent syncope as persistently isolated feature of transthyretin amyloidotic polyneuropathy. Neuromuscul Disord 15:259–261
15. Klein AL, Tajik AJ (1991) Doppler assessment of pulmonary venous flow in healthy subjects and in patients with heart disease. J Am Soc Echocardiogr 4:379–392
16. Bellavia D, Pellikka PA, Abraham TP et al (2008) Evidence of impaired left ventricular systolic function by Doppler myocardial imaging in patients with systemic amyloidosis and no evidence of cardiac involvement by standard two-dimensional and Doppler echocardiography. Am J Cardiol 101:1039–1045
17. Lebovic D, Hoffman J, Levine BM et al (2008) Predictors of survival in patients with systemic light-chain amyloidosis and cardiac involvement initially ineligible for stem cell transplantation and treated with oral melphalan and dexamethasone. Br J Haematol 143(3):369–373
18. Gertz MA (2008) Troponin in hematologic oncology. Leuk Lymphoma 49:194–203
19. Dispenzieri A, Gertz MA, Kyle RA et al (2004) Serum cardiac troponins and N-terminal pro-brain natriuretic peptide: a staging system for primary systemic amyloidosis. J Clin Oncol 22:3751–3757
20. Dispenzieri A, Gertz MA, Kyle RA et al (2004) Prognostication of survival using cardiac troponins and N-terminal pro-brain natriuretic peptide in patients with primary systemic amyloidosis undergoing peripheral blood stem cell transplantation. Blood 104:1881–1887
21. Gertz M, Lacy M, Dispenzieri A, Hayman S, Kumar S, Buadi F, Leung N, Litzow M (2008) Troponin T level as an exclusion criterion for stem cell transplantation in light-chain amyloidosis. Leuk Lymphoma 49:36–41
22. Gertz MA, Kyle RA (1989) Primary systemic amyloidosis–a diagnostic primer. Mayo Clin Proc 64:1505–1519
23. Gertz MA, Lacy MQ, Dispenzieri A, Hayman SR (2005) Amyloidosis: diagnosis and management. Clin Lymphoma Myeloma 6:208–219
24. Kyle RA, Gertz MA (1995) Primary systemic amyloidosis: clinical and laboratory features in 474 cases. Semin Hematol 32:45–59
25. Palladini G, Kyle RA, Larson DR, Therneau TM, Merlini G, Gertz MA (2005) Multicentre versus single centre approach to rare diseases: the model of systemic light chain amyloidosis. Amyloid 12:120–126
26. Sanchorawala V, Seldin DC, Magnani B, Skinner M, Wright DG (2005) Serum free light-chain responses after high-dose intravenous melphalan and autologous stem cell transplantation for AL (primary) amyloidosis. Bone Marrow Transplant 36:597–600
27. Lachmann HJ, Gallimore R, Gillmore JD et al (2003) Outcome in systemic AL amyloidosis in relation to changes in concentration of circulating free immunoglobulin light chains following chemotherapy. Br J Haematol 122:78–84
28. Skinner M, Sanchorawala V, Seldin DC et al (2004) High-dose melphalan and autologous stem-cell transplantation in patients with AL amyloidosis: an 8-year study. Ann Intern Med 140:85–93
29. Gertz MA, Lacy MQ, Dispenzieri A et al (2007) Effect of hematologic response on outcome of patients undergoing transplantation for primary amyloidosis: importance of achieving a complete response. Haematologica 92:1415–1418
30. Goodman HJ, Gillmore JD, Lachmann HJ, Wechalekar AD, Bradwell AR, Hawkins PN (2006) Outcome of autologous stem cell transplantation for AL amyloidosis in the UK. Br J Haematol 134:417–425

31. Wiseman RL, Powers ET, Kelly JW (2005) Partitioning conformational intermediates between competing refolding and aggregation pathways: insights into transthyretin amyloid disease. Biochemistry 44:16612–16623
32. Gertz MA, Lacy MQ, Dispenzieri A, Hayman SR, Kumar S (2007) Transplantation for amyloidosis. Curr Opin Oncol 19:136–141
33. Palladini G, Perfetti V, Obici L et al (2004) Association of melphalan and high-dose dexamethasone is effective and well tolerated in patients with AL (primary) amyloidosis who are ineligible for stem cell transplantation. Blood 103:2936–2938
34. Jaccard A, Moreau P, Leblond V et al (2007) High-dose melphalan versus melphalan plus dexamethasone for AL amyloidosis. N Engl J Med 357:1083–1093
35. Gertz MA, Kyle RA (1990) Acute leukemia and cytogenetic abnormalities complicating melphalan treatment of primary systemic amyloidosis. Arch Intern Med 150:629–633
36. Gertz MA, Lacy MQ, Lust JA, Greipp PR, Witzig TE, Kyle RA (2008) Long-term risk of myelodysplasia in melphalan-treated patients with immunoglobulin light-chain amyloidosis. Haematologica 93(9):1402–1406
37. Palumbo A, Bringhen S, Liberati AM et al (2008) Oral melphalan, prednisone, and thalidomide in elderly patients with multiple myeloma: updated results of a randomized, controlled trial. Blood 112(8):3107–3114
38. Mateos MV, Hernandez JM, Hernandez MT et al (2006) Bortezomib plus melphalan and prednisone in elderly untreated patients with multiple myeloma: results of a multicenter phase 1/2 study. Blood 108:2165–2172
39. Kastritis E, Dimopoulos MA (2007) The evolving role of lenalidomide in the treatment of hematologic malignancies. Expert Opin Pharmacother 8:497–509
40. Palladini G, Perfetti V, Perlini S et al (2005) The combination of thalidomide and intermediate-dose dexamethasone is an effective but toxic treatment for patients with primary amyloidosis (AL). Blood 105:2949–2951
41. Cohen AD, Zhou P, Chou J et al (2007) Risk-adapted autologous stem cell transplantation with adjuvant dexamethasone +/− thalidomide for systemic light-chain amyloidosis: results of a phase II trial. Br J Haematol 139:224–233
42. Wechalekar AD, Gillmore JD, Lachmann HJ, Offer M, Hawkins PN (2006) Efficacy and safety of bortezomib in systemic AL amyloidosis – a preliminary report (abstract). Blood 108:129
43. SV RDE, Hegenbart U, Merlini G, Palladini G, Fermand J, Vescio RA, Liu X, Elsayed YA, Comenzo RL (2007) Phase I/II study of bortezomib (B) in patients with systemic AL-amyloidosis (AL). JCO 25:453s
44. Dispenzieri A, Lacy MQ, Zeldenrust SR et al (2007) The activity of lenalidomide with or without dexamethasone in patients with primary systemic amyloidosis. Blood 109:465–470
45. Sanchorawala V, Wright DG, Rosenzweig M et al (2007) Lenalidomide and dexamethasone in the treatment of AL amyloidosis: results of a phase 2 trial. Blood 109:492–496
46. Takayama S, Reed JC, Homma S (2003) Heat-shock proteins as regulators of apoptosis. Oncogene 22:9041–9047
47. Sitia R, Palladini G, Merlini G (2007) Bortezomib in the treatment of AL amyloidosis: targeted therapy? Haematologica 92:1302–1307
48. Zhang N, Ahsan MH, Zhu L, Sambucetti LC, Purchio AF, West DB (2005) Regulation of IkappaBalpha expression involves both NF-kappaB and the MAP kinase signaling pathways. J Inflamm (Lond) 2:10
49. Anderson KC, Alsina M, Bensinger W et al (2007) Multiple myeloma. Clinical practice guidelines in oncology. J Natl Compr Canc Netw 5:118–147
50. Wechalekar AD, Lachmann HJ, Offer M, Hawkins PN, Gillmore JD (2008) Efficacy of bortezomib in systemic AL amyloidosis with relapsed/refractory clonal disease. Haematologica 93:295–298
51. Fauble VS, Shah-Reddy I (2006) Primary amyloidosis treated with bortezomib with a clinical and radiological response (abstract 5111). Blood 108

52. Kastritis E, Anagnostopoulos A, Roussou M et al (2007) Treatment of light chain (AL) amyloidosis with the combination of bortezomib and dexamethasone. Haematologica 92:1351–1358
53. Roussou M, Kastritis E, Migkou M et al (2008) Treatment of patients with multiple myeloma complicated by renal failure with bortezomib-based regimens. Leuk Lymphoma 49:890–895
54. Leung N, Dispenzieri A, Fervenza FC et al (2005) Renal response after high-dose melphalan and stem cell transplantation is a favorable marker in patients with primary systemic amyloidosis. Am J Kidney Dis 46:270–277

Second-Generation Proteasome Inhibitors

Dixie-Lee Esseltine, Larry Dick, Erik Kupperman, Mark Williamson, and Kenneth C. Anderson

Abstract The first-in-class proteasome inhibitor, bortezomib, has provided proof-of-concept for the therapeutic approach of proteasome inhibition in a number of malignancies. However, as we look to the future and to further improving upon the contributions of this class of drugs, we will need to consider optimizing activity in solid tumors, reducing peripheral neuropathy and utilizing more convenient routes of administration. A number of "second-generation" proteasome inhibitors have been identified and are now in preclinical and clinical development, including MLN9708, CEP-18770, carfilzomib, and salinosporamide A (NPI-0052). These agents differ from bortezomib in some of their key characteristics, and differences in their pharmacology may result in different activity and safety profiles. This chapter reviews the second-generation proteasome inhibitors, together with other potential therapeutic targets in the ubiquitin–proteasome system.

1 Introduction: Bortezomib and Beyond

With the development and regulatory approval of bortezomib (VELCADE®), proteasome inhibition emerged as a feasible and effective therapeutic approach for developing drugs, potentially for a number of malignancies [1, 2]. Bortezomib was the first proteasome inhibitor introduced to the clinic and has provided proof of concept for this approach [3]. A variety of other molecules that have been developed to exploit this novel mechanism are currently in preclinical or clinical

D.-L. Esseltine (✉), L. Dick, E. Kupperman, and M. Williamson
Millennium Pharmaceuticals Inc, Cambridge, MA, USA
e-mail: Dixie-Lee.Esseltine@mpi.com, larry.dick@mpi.com, Erik.Kupperman@mpi.com, mark.williamson@mpi.com

K.C. Anderson
Jerome Lipper Center for Multiple Myeloma, Dana-Farber Cancer Institute, Boston, MA, USA
e-mail: Kenneth_Anderson@dfci.harvard.edu

I.M. Ghobrial et al. (eds.), *Bortezomib in the Treatment of Multiple Myeloma*,
Milestones in Drug Therapy,
DOI 10.1007/978-3-7643-8948-2_11,

development, with the aim of building upon the established success of bortezomib. In this chapter, we examine the development of "second-generation" proteasome inhibitors, and compare and contrast their key characteristics with those of bortezomib, the "first-generation" proteasome inhibitor. It is important to note that the second-generation inhibitors are at relatively early stages in development compared with bortezomib; therefore, it is not possible at this point to determine whether the conventional paradigm of second-generation drugs improving upon first-generation agents will hold true with proteasome inhibitors. Only clinical trials will determine whether the pharmacological differences between the second- and first-generation agents will translate into important differences in safety and efficacy in the clinic.

1.1 Inhibition of the 26S Proteasome

The identification of the proteasome as a druggable target became possible as a result of our increased understanding over the past 3 decades of the ubiquitin–proteasome system and the crucial role that the 26S proteasome plays in the regulation and degradation of intracellular proteins [4–7]. Many substrates of the 26S proteasome are proteins that regulate cellular processes known to be important for normal cell growth, such as those regulating the cell cycle, growth signaling, and levels of pro and antiapoptotic proteins. Dysregulation of any of these cellular processes can lead to malignant transformation in cells, resulting in the development and growth of tumors; therefore, the ubiquitin–proteasome system emerged as a unique target for anticancer therapy [1, 2, 8].

Prior to the development of bortezomib, various compounds have been identified as inhibitors of the proteasome, including agents in the following five main classes: peptide aldehydes, peptide vinyl sulfones, peptide boronates, peptide epoxyketones (epoxomycin and eponomycin), and β-lactones (lactacystin and derivatives) [2, 9, 10]. However, most of these early compounds were not suitable for clinical development because of insufficient potency, metabolic instability, or lack of specificity for the proteasome. Bortezomib (Fig. 1) was selected for clinical development from a series of dipeptide boronic acid analogs, based on its high potency for inhibiting proteasome

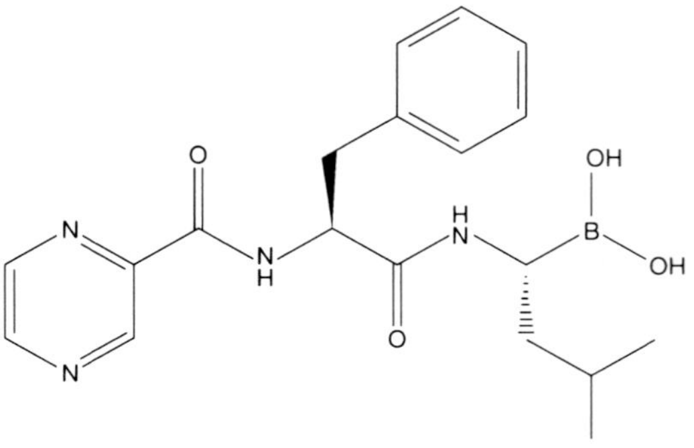

Fig. 1 Bortezomib, a dipeptide boronic acid analog [13]

activity, the reversible nature of this inhibition, and its specificity for inhibiting the 26S proteasome [11, 12].

The 26S proteasome comprises a 20S core particle with a 19S cap at either end. Within the 20S core are three different active enzymatic sites, with trypsin-like, chymotrypsin-like, and post-glutamyl peptide hydrolase-like (caspase-like) activities (Fig. 2) [14, 15].

Bortezomib specifically and reversibly inhibits the chymotrypsin-like activity of the 20S proteasome, as shown in Fig. 2; it also inhibits the caspase-like activity, albeit to a lesser extent. This inhibition occurs through interactions with threonine residue on these β subunits (Fig. 3) [2].

1.2 Preclinical Activity of Bortezomib

Preclinical studies with bortezomib have provided proof of the putative mechanisms of action of proteasome inhibition in a range of tumor types. Bortezomib has been shown to inhibit nuclear factor (NF)-κB activity by preventing degradation of its inhibitor IκB, to deregulate the turnover of cyclins, thereby disrupting cyclin-dependent kinase activity, to stabilize the tumor suppressor p53, and to shift the proapoptotic/antiapoptotic balance in the Bcl-2 family of proteins, among other effects [1, 3, 8]. In addition, the antitumor activity of bortezomib has been linked to the accumulation of misfolded proteins in cells, causing endoplasmic reticulum stress and triggering the unfolded protein response, ultimately resulting in cell death [17, 18]. Furthermore, preclinical studies have shown that these mechanisms result in additive or synergistic activity between proteasome inhibition with bortezomib and the cytotoxic or targeted effects of conventional or novel agents [19–30].

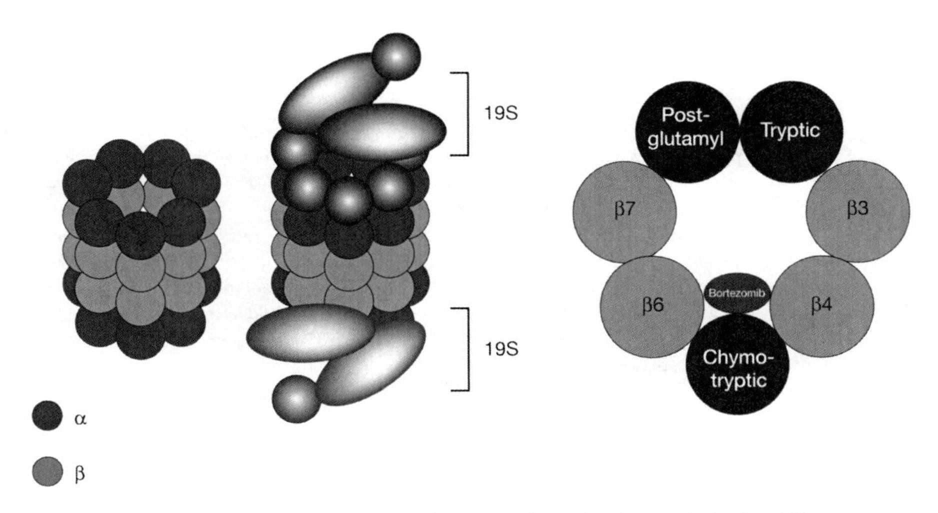

Fig. 2 Structure of the 26S proteasome and cross-section of active catalytic sites [2]

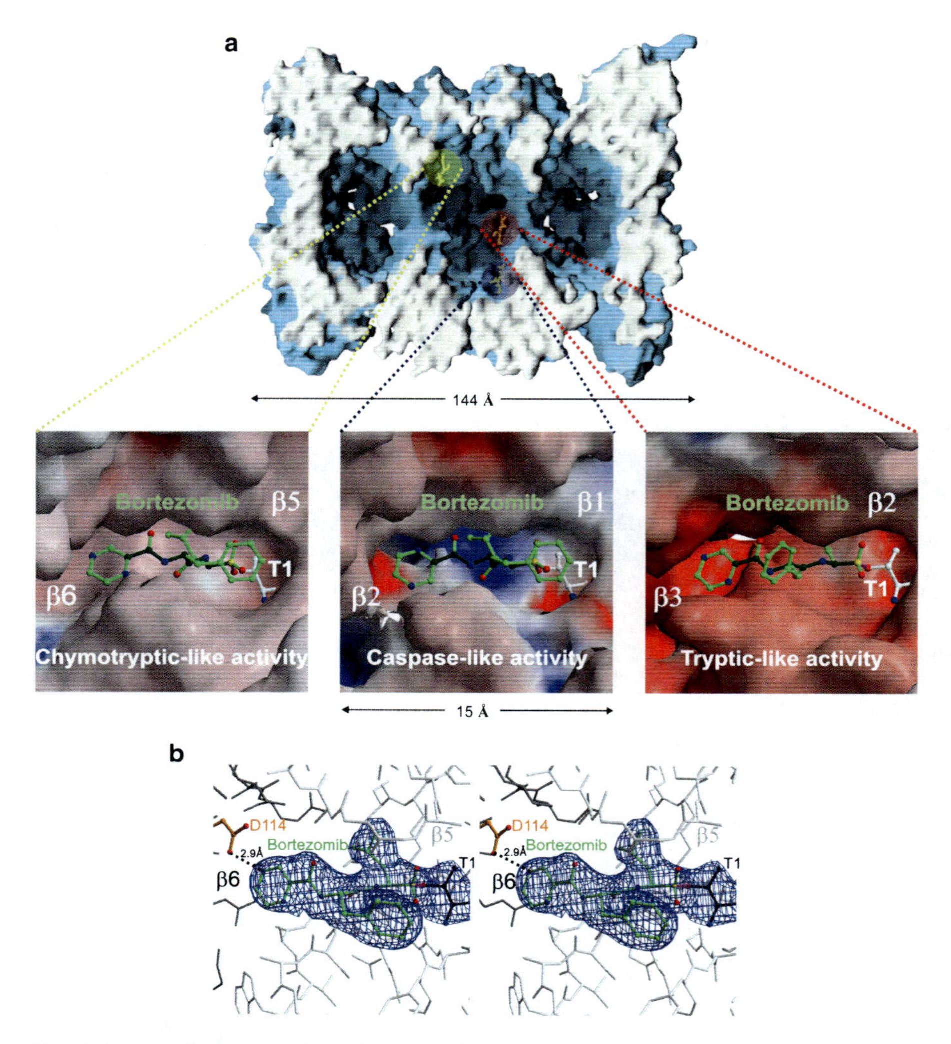

Fig. 3 Structural representation of bortezomib bound to distinct active sites in the yeast 20S proteasome. (**a**) Surface representation of the yeast 20S proteasome (*blue*) crystallized in the presence of bortezomib, clipped along the cylindrical pseudo-sevenfold symmetry axis (*white*). The various proteolytic surfaces are marked by a specific color coding: *red* = subunit β1; *blue* = subunit β2; and *green* = subunit β5. Cleavage preferences, termed chymotryptic-like, caspase-like, and tryptic-like activity, are zoomed and illustrated in surface representations; the nucleophilic threonine and bortezomib are presented as a ball-and-stick model. Basic residues are colored blue, acidic residues are red, and hydrophobic residues are white. (**b**) Stereorepresentation of the chymotryptic-like active site of the yeast 20S proteasome (*white* and *gray*) and bortezomib (*green*). Covalent linkage of the inhibitor with the active site Thr1 of subunit β5 is drawn in *pink*. The electron density map (*blue*) is contoured from 1.2σ, with $2F_o–F_c$ coefficients after twofold averaging. Apart from the bound inhibitor, structural changes were noted only in the specificity pockets. Temperature factor refinement indicates full occupancy of bortezomib bound to the chymotryptic-like active site. Bortezomib has been omitted for phasing. (Reproduced with permission from Groll et al. [16], Structure 2006;14:451-456)

1.3 Clinical Activity of Bortezomib

In the clinic, these mechanisms have translated into substantial activity with bortezomib alone or in combination in various hematologic malignancies, including, most notably, multiple myeloma (both in previously untreated patients [17, 31] and in those with relapsed and/or refractory disease [32, 33]), and mantle cell lymphoma (again, in both the front-line [34–36] and relapsed/refractory [36–40] settings); bortezomib alone and in combination has also demonstrated activity in relapsed/refractory follicular and other indolent lymphomas [41–44], previously untreated and relapsed/refractory Waldenström's macroglobulinemia [45–48], and primary systemic amyloidosis [49, 50]. Consequently, at the time of publication, bortezomib is approved in the US for the treatment of patients with multiple myeloma, and for the treatment of patients with mantle cell lymphoma who have received at least one prior therapy [13]. In the European Union, bortezomib is approved for the treatment of previously untreated multiple myeloma in combination with melphalan and prednisone, and single-agent bortezomib is approved for the treatment of multiple myeloma following at least one prior therapy in countries worldwide.

1.4 Limitations of Bortezomib

Despite the success of bortezomib, there are aspects of its activity, safety profile, and administration that are areas for refinement, namely the lower-than-desirable activity in solid tumors, the development of peripheral neuropathy in approximately one-third of patients, and the intravenous route of delivery; in addition, cellular resistance to bortezomib has been demonstrated [51–54], and in the clinic some patients may not be sensitive or may become refractory to bortezomib treatment. Clinical studies of bortezomib in solid tumors, notably prostate cancer [55, 56] and non-small cell lung cancer [57–59], have not reflected activity seen in preclinical investigations in cell lines of these tumor types [12, 60–64] as strongly as with the hematologic malignancies; this may be a result of the pharmacological profile of bortezomib, with high availability in the blood but limited distribution into solid tumors. Peripheral neuropathy is a known toxicity associated with bortezomib treatment [65, 66]. Studies in vitro and in animal models suggest that this neurotoxicity is a class effect associated with proteasome inhibition, with the accumulation of neuronal cytoplasmic aggregates having been demonstrated with proteasome inhibitors of different chemical scaffolds [67–69]. More clinical experience with other proteasome inhibitors is required to determine whether this is borne out in patients. Finally, bortezomib is currently administered by standard twice-weekly or weekly intravenous (IV) injection, requiring patient attendance at the tertiary care center, and this may be inconvenient for some patients [13]; a pivotal study is ongoing for further validation of the subcutaneous route of administration, which could be undertaken at a patient's local clinic [70]. It is also clear that oral administration may improve the convenience of therapy. Second-generation agents need to address these limitations of bortezomib while retaining its substantial existing efficacy. It is only then that such second- or later-generation agents will truly represent an important improvement in this therapeutic class.

2 Second-Generation Inhibitors of the 26S Proteasome

As noted earlier, a number of "second-generation" proteasome inhibitors have been identified and are in preclinical and clinical development by various pharmaceutical companies, including Millennium Pharmaceuticals, Inc. These agents differ from bortezomib in some of their key characteristics, including their chemical class, the active sites of the 20S proteasome that they inhibit, and the reversibility of this inhibition, as shown in Table 1. Differences in the pharmacology of the compounds may result in different activity and safety profiles, with some preclinical studies suggesting altered antitumor activity. The clinical dosing schedules being investigated for the second-generation agents may also differ from those for bortezomib as a consequence.

The key question with regard to the second-generation proteasome inhibitors is whether the chemical and pharmacologic differences between them and bortezomib will result in their proving to be better drugs in the clinic. As noted earlier, it is too early to answer this because these agents are currently only in their early stages of clinical development. However, it is instructive to speculate on the potential impact of the differences in pharmacology and to review the available preclinical data on the second-generation agents.

2.1 *Mechanisms of Activity*

First, it should be noted that bortezomib and the second-generation proteasome inhibitors CEP-18770 (Cephalon Oncology), carfilzomib (PR-171; Proteolix, Inc.), and salinosporamide A (NPI-0052; Nereus Pharmaceuticals, Inc.) all have the same basic mechanism of activity, that is, inhibition of 20S proteasome activity through the blocking of one or more of the β subunit catalytic sites. Furthermore, preclinical studies suggest that each of the compounds is an equally highly effective inhibitor of 20S proteasome activity.

For example, Williamson et al. showed that bortezomib and ML858 (synthetically accessible salinosporamide A) were equipotent in terms of inactivation of the chymotrypsin-like site (β5 subunit) of the 20S proteasome and inhibition of NF-κB activity in vitro, with IC_{50} values of 40 and 20 nmol/L, respectively, for NF-κB inhibition [83]. The compounds differed somewhat in their relative selectivity for the caspase-like (β1 subunit) and trypsin-like (β2 subunit) sites, but differences were minor [83]. In addition, bortezomib and CEP-18770 demonstrated comparable

Table 1 Characteristics of bortezomib and the second-generation proteasome inhibitors

	Bortezomib	CEP-18770	Carfilzomib (PR-171)	Salinosporamide A (NPI-0052)
Structure	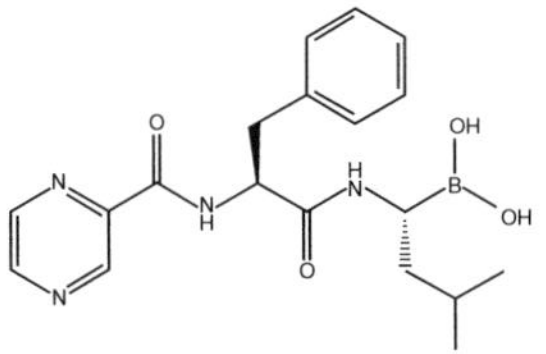	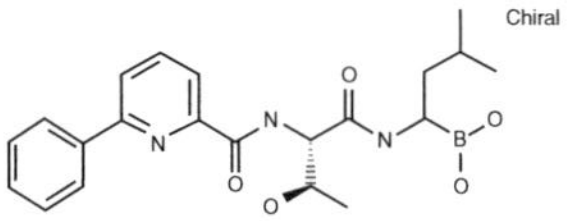	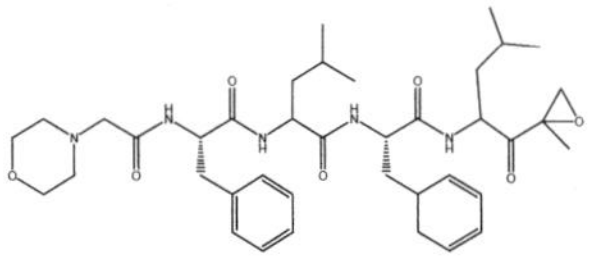	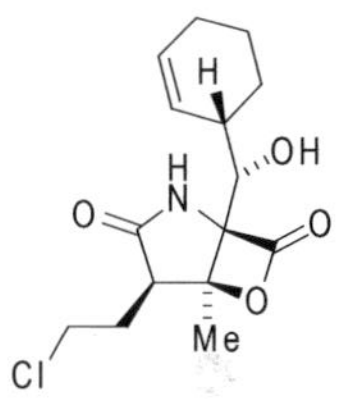
Class	Peptide boronic acid [13] (dipeptide boronic acid analog)	Peptide boronic acid [71, 72] (P2 threonine boronic acid)	Peptide epoxyketone [73] (epoxomicin analog)	β-lactone [74] (non-peptide bicyclic γ-lactam–β-lactone resembling omuralide, the active form of lactacystin; derived from the marine bacterium *Salinospora tropica* [75])
Proteasome inhibition	Reversible	Reversible	Irreversible [73]	Irreversible [76]
Sites	Primarily β5 (chymotrypsin-like) [77]	Primarily β5 [71, 72]	Primarily β5 [73]	Primarily β5, plus β2 (trypsin-like), β1 (caspase-like) [76, 78–81]
Route of administration	IV (sc route studied in clinic [70])	IV (potential for oral admin)	IV (oral analog PR-047 in development [82])	IV (potential for oral admin)

potency in vitro, with similar low-nanomolar IC_{50} values for inhibition of the chymotrypsin-like and caspase-like activities in multiple myeloma lysates [72]; comparable inhibition of NF-κB activity in RPMI-8226 and U266 multiple myeloma cells was also seen [72]. Finally, Demo et al. also reported comparable potency in terms of inhibition of chymotrypsin-like proteasome activity among bortezomib, carfilzomib, and salinosporamide A, although they showed carfilzomib to be less potent at inhibiting the caspase-like and trypsin-like activities when compared with bortezomib and salinosporamide A [73], suggesting greater selectivity [73] for the chymotrypsin-like site and a lower impact on overall protein turnover [84]. The clinical relevance of this finding is not yet clear, but the authors suggested that it may have contributed to the improved tolerability observed in in vivo studies in rodents, in terms of no changes in hematocrit or hemoglobin and no significant weight loss being noted in animals receiving repeated administration of carfilzomib for up to five consecutive days [73]. Conversely, it may be hypothesized that a reduced impact on overall protein turnover may result in a more limited accumulation of misfolded proteins as a result of proteasome inhibition, thereby resulting in less endoplasmic reticulum stress and reducing the subsequent unfolded protein response and ensuing cell death. Related to this, on the basis of enzyme studies, it has also been suggested that bortezomib but not carfilzomib may inhibit non-proteasomal enzymes; however, the relevance of this finding regarding off-target effects in terms of the clinical toxicity profile remains to be determined [85].

Nevertheless, the cellular mechanisms of activity of bortezomib and the second-generation proteasome inhibitors appear similar. Bortezomib has been shown to cause caspase-mediated apoptosis via the caspase-9-dependent intrinsic mitochondrial pathway, the caspase-8-mediated extrinsic death receptor pathway, and, as noted earlier, the endoplasmic reticulum stress response pathway, involving caspase-12 [20, 21, 86–90]. Associated effects include activation of c-Jun NH_2-terminal kinase (JNK) and increased c-Jun phosphorylation, reduced levels of cellular FLICE-like inhibitory protein (c-FLIP, a negative regulator of Fas-induced apoptosis), cleavage of poly(ADP-ribose) polymerase (PARP), downregulation of the insulin-like growth factor-1 (IGF-1) signaling pathway, p42/44 mitogen-activated protein kinase (MAPK) inhibition, and, as discussed earlier, inhibition of NF-κB activity [8, 20, 21, 86–89]. Similar effects have been reported in cellular studies of the second-generation proteasome inhibitors. NPI-0052 has been shown to cause caspase-8-mediated death signaling [78, 91], inhibition of NF-κB activity [80, 92–95], reduction in the expression of cFLIP and other antiapoptotic proteins [92], induction of Raf-1 kinase inhibitor protein [96], upregulation of death receptor 5 expression [95], and PARP cleavage [93], as well as endoplasmic reticulum stress-mediated apoptosis [79]. CEP-18770 also inhibits NF-κB activity, and results in caspase activation and associated PARP cleavage [72]. Carfilzomib has been shown to cause apoptosis in hematologic tumor cell lines through both the intrinsic and extrinsic caspase pathways [97], associated with an increase in activated JNK, PARP cleavage, and accumulation of markers of apoptotic, growth arrest, and stress response pathways [73, 97].

2.2 *Reversible vs. Irreversible Proteasome Inhibition*

One of the notable differences between the various proteasome inhibitors is the reversibility of their covalent adduct binding to the β subunits of the 20S proteasome resulting in inhibition. Bortezomib and CEP-18770 [72] are both reversible inhibitors, whereas with carfilzomib [73] and salinosporamide A [76], the inhibition is irreversible. This apparent difference may be obscured in the biologic setting, however, since any kind of proteasome inhibition may be considered reversible in nucleated cells, because of their capacity to synthesize new proteasomes.

The pharmacodynamic profile of the reversible proteasome inhibitors differs in some respects from that of the irreversible inhibitors. For example, sustained whole blood proteasome inhibition is seen with both the irreversible inhibitors carfilzomib [73] and NPI-0052 [98], whereas with bortezomib and CEP-18770, recovery of whole blood proteasome function is seen within 48–72 h of administration [72]. Following intravenous administration of carfilzomib in rats and mice, less than 50% recovery of 20S proteasome activity was observed in whole blood after 1 week [73], and similarly, single intravenous administration of NPI-0052 resulted in sustained whole blood proteasome inhibition with significant recovery by day 7 [98]. This is due to the fact that anucleated cells such as red blood cells are not able to synthesize new proteasomes and have a lengthy half-life of approximately 120 days. This long half-life results in a lengthy pharmacodynamic effect from irreversible covalent inhibition of the 20S proteasome in these cells [73, 98]. By contrast, in nucleated cells such as peripheral blood mononuclear cells, which can generate new proteasomes and have a shorter half-life of 2–5 days [98], proteasome activity recovered within 48–72 h following NPI-0052 administration [98], while proteasome activity recovered by 50–100% within 24 h in all tissues examined following carfilzomib administration [73]. Such recovery periods were similar to those seen with the slowly reversible binding of bortezomib [73] and CEP-18770 [72], indicating that in tissues other than whole blood, the synthesis of new proteasomes is the major determinant of the recovery of 20S proteasome activity with all the proteasome inhibitors [73].

These findings provide an insight into some of the differences observed between bortezomib and the irreversible second-generation proteasome inhibitors in in vitro and in vivo studies. For example, in vitro studies in various cell lines have suggested that carfilzomib has greater cytotoxic activity compared with bortezomib following brief pulsed exposure of cells to either agent, but that this differential is eliminated upon prolonged exposure [73, 97, 99]. In studies reported by Demo et al., myeloma, lymphoma, and various solid tumor cell lines were exposed to carfilzomib or bortezomib for 1 h, followed by a 72-h washout. Unsurprisingly given the irreversible nature of the covalent binding with carfilzomib, cell viability at the end of washout was lower with carfilzomib compared with bortezomib; however, if cells were exposed to either agent for 72 h, cell viability was similarly low with carfilzomib and bortezomib, indicating comparable cytotoxic activity [73]. Kuhn et al. also showed that 1-h pulsed exposure of various cell lines and cells derived from patients with diffuse large B cell lymphoma, chronic lymphocytic leukemia, and acute myeloid leukemia to carfilzomib, followed by recovery over 24 h, appeared to result in greater activity compared with bortezomib [97]. Although such in vitro exposure was designed to mimic the in vivo pharmacokinetic profile of proteasome inhibition, only clinical studies will be able to determine whether such findings are reflected as differences in activity in patients. Similarly, in a colorectal cancer mouse xenograft model, Cusack et al. reported greater whole blood proteasome inhibition at 1.5 h following intravenous administration of the maximum tolerated dose of NPI-0052 compared with bortezomib [94]; however, as noted in human studies, peak proteasome inhibition with bortezomib occurs within minutes of intravenous administration [70, 100], with the reversible nature of this inhibition subsequently resulting in recovery of proteasome activity over the following 48–72 h. By contrast, irreversible inhibition with NPI-0052 results in sustained whole blood proteasome inhibition, as would be expected.

Although irreversible proteasome inhibitors appear to offer notable cytotoxic activity in preclinical studies, it is possible that the pharmacology of such agents may result in reduced distribution into solid tumors in humans when compared with reversible proteasome inhibitors. A greater majority of these molecules, compared with reversible inhibitors, may end up irreversibly bound to proteasomes in red blood cells and hepatocytes, and may therefore not be able to access solid tumor tissue. The blood could in effect act as a very large "sink" for these agents, potentially limiting efficacy in solid tumors, even to a greater extent than that seen with the slowly reversible proteasome inhibition of bortezomib. More extensive clinical studies are required especially in patients with solid tumors to determine the activity of the second-generation proteasome inhibitors and the effect of irreversible binding.

2.3 *Activity of Second-Generation Proteasome Inhibitors in Preclinical Studies*

Regardless of the differences in the characteristics discussed earlier, the second-generation proteasome inhibitors have demonstrated substantial activity alone and in combination in preclinical studies, reflecting a similar preclinical experience to bortezomib.

2.3.1 Single-Agent Activity

Like bortezomib [20, 21, 86, 88], CEP-18770 has been shown to induce apoptosis in multiple myeloma cell lines and in cells from multiple myeloma patients [72], resulting in complete tumor regressions and median survival benefits in multiple myeloma mouse xenograft models [72]. Carfilzomib also demonstrated dose-dependent inhibition of proliferation, resulting in apoptosis in multiple myeloma cell lines and in patients' cells, and showed activity in dexamethasone- and melphalan-resistant, but not doxorubicin-resistant multiple myeloma cells [97]. Single-dose pharmacokinetic studies in rats and monkeys demonstrated dose-linear exposure of carfilzomib [101], while prolonged exposure to carfilzomib appeared well tolerated in rats and monkeys, with no impact on neurobehavioral function [102]. Dose- and schedule-dependent antitumor efficacy was seen in human tumor xenograft models of colorectal adenocarcinoma, B cell lymphoma, and Burkitt's lymphoma [73]. In these mouse models, carfilzomib administered on 2 consecutive days on a weekly schedule appeared more efficacious than bortezomib on a twice-weekly schedule [73]; however, it remains to be seen whether these preclinical data will translate into differences in clinical activity in the respective tumor types. Extensive preclinical experience with NPI-0052 has also demonstrated similar antitumor activity to bortezomib and the other second-generation proteasome inhibitors. NPI-0052 was shown to be active in multiple myeloma cells, including those resistant to conventional therapies, and has demonstrated anti-myeloma activity and prolonged survival in animal models of multiple myeloma [81, 91]. It has also demonstrated activity in other hematologic tumor cell lines, including Hodgkin lymphoma, mantle cell lymphoma and other non-Hodgkin's lymphoma, Waldenström's macroglobulinemia, and leukemia cells [78, 80, 103, 104], and in vivo in leukemic mice [78].

2.3.2 Activity in Bortezomib-Resistant Cells

Of potential importance is the activity seen with second-generation proteasome inhibitors in bortezomib-resistant cells and in cells from patients who have either relapsed following bortezomib treatment or demonstrated clinical bortezomib resistance. CEP-18770 resulted in multiple myeloma cell death in primary multiple myeloma biopsies cultured ex vivo from three bortezomib-resistant patients, although bortezomib also appeared to offer similar activity in this assay [72], while NPI-0052 was also active in multiple myeloma cells resistant to bortezomib [91]. Similarly, bortezomib-resistant cell lines and cells from patients who had clinical progression on bortezomib demonstrated greater sensitivity to carfilzomib than bortezomib, albeit with some degree of cross-resistance in the bortezomib-resistant cell lines, and carfilzomib was shown to inhibit proliferation of plasma cells from patients who had no benefit from bortezomib therapy [97]. It remains to be seen if these in vitro and ex vivo findings are reflected in clinical activity in bortezomib-refractory patients.

2.3.3 Enhanced Activity in Combination

Preclinical studies in various tumor cell lines have demonstrated that bortezomib has additive or synergistic cytotoxic activity with a range of conventional therapeutic agents, such as dexamethasone, doxorubicin, and melphalan [19–22], and novel targeted agents, including histone deacetylase inhibitors [23, 24], pan-Bcl2 inhibitors [25–27], and tumor necrosis factor-alpha-related apoptosis-inducing ligand (TRAIL) [28–30]. Increased activity of these combinations seems to be attributable to the multiple distinct mechanisms of activity of proteasome inhibition. As expected, second-generation proteasome inhibitors also demonstrate synergistic or additive activity when combined with conventional and novel therapeutic agents in preclinical studies. CEP-18770 resulted in synergistic anti-myeloma activity in combination with melphalan or doxorubicin in vitro in myeloma cell lines and with melphalan in a mouse model [105]. Carfilzomib showed synergistic activity in combination with dexamethasone in multiple myeloma cells [97] and in combination with the cyclin-dependent kinase-4 and -6 inhibitor PD 0332991 in chemoresistant myeloma cells [106], as well as synergistic apoptosis in the mantle cell lymphoma cell lines HBL-2 and Granta in combination with the BH3 mimetic AT-101 [107]. In immunocompromised mice, it also resulted in significant reductions in tumor growth in combination with docetaxel in an A549 lung carcinoma model and with liposomal doxorubicin in a HT-29 colorectal carcinoma model, compared with individual agents alone [108]. Similarly, NPI-0052 demonstrated synergistic anti-myeloma activity in vitro in multiple myeloma cells and significant inhibition of tumor growth and prolongation of survival in a mouse myeloma model in combination with lenalidomide [109], showed synergistic apoptosis in combination with histone deacetylase inhibitors in leukemia cells [78, 110], enhanced the cytotoxicity of tumor necrosis factor and thalidomide in KBM5 myeloid leukemia cells in vitro [92], sensitized B cell non-Hodgkin's lymphoma cells to TRAIL [95, 96, 104], and showed enhanced activity in mantle cell lymphoma cells in combination with the pan-Bcl-2 inhibitor obatoclax [111]. In solid tumor cell lines, NPI-0052 enhanced apoptosis in combination with SN-38/oxaliplatin, 5-FU, and leucovorin in colorectal cancer cells [94], while low-dose combinations of NPI-0052 and TRAIL caused marked cell death in various pancreatic cancer cells [112]. NPI-0052 plus the histone deacetylase inhibitor vorinostat also resulted in enhanced apoptosis in vitro in pancreatic cancer cells and, in vivo, demonstrated decreased tumor growth in mice [113]. Significantly enhanced tumor growth response was seen when NPI-0052 was added to the triplet combination of gemcitabine, erlotinib, and bevacizumab in a pancreatic cancer xenograft model [114]. In addition, NPI-0052 was shown to sensitize TRAIL-resistant prostate cancer cells to TRAIL through NF-κB inhibition [93].

Interestingly, NPI-0052 has demonstrated synergistic activity in combination with bortezomib in vitro [91] and in vivo [115] in multiple myeloma and in Waldenström's macroglobulinemia cells [80], and enhanced the cytotoxicity of bortezomib in KBM5 myeloid leukemia cells in vitro [92]. These findings suggest that bortezomib and NPI-0052 have differing effects on cellular signaling pathways and on the caspase-like and trypsin-like activities of the 20S proteasome, and thereby have different but complementary mechanisms of action in multiple myeloma and Waldenström's macroglobulinemia cells [80, 91]. Alternatively, these effects may be associated with bortezomib and NPI-0052 competing for 20S proteasome binding sites, altering the distribution of the agents into tissues [83].

2.4 Clinical Studies of Second-Generation Proteasome Inhibitors

Data have been reported from a number of phase 1 and 2 studies with the second-generation proteasome inhibitors. At the time of writing this chapter, no clinical data had been presented on CEP-18770, but results were available from three phase 1 studies of NPI-0052 in patients with relapsed and relapsed/refractory multiple myeloma, lymphoma and solid tumors, and advanced malignancies [116–118]. In the study in multiple myeloma, ten patients with relapsed or relapsed/refractory myeloma have been treated with NPI-0052 at doses of 0.025–0.075 mg/m^2 on days 1, 8, and 15 of 4-week cycles; the maximum tolerated dose has not been reached. No responses have been confirmed, but two patients remained on study for 6 months and 1 year with stable disease [116]. In another study, a total of 29 patients have been treated with escalating NPI-0052 doses of 0.0125–0.375 mg/m^2, with no maximum tolerated dose being reached. A dose–pharmacodynamic response relationship was reported in this and other clinical trials with NPI-0052, with proteasome inhibition of up to 92%. No responses were seen, but seven patients demonstrated stable disease lasting more than 2 months, including patients with melanoma, colorectal carcinoma, hepatocellular carcinoma, adenoid cystic carcinoma, ovarian carcinoma, and cervical carcinoma [117]. In the third phase 1 study, 22 patients have been treated with NPI-0052 doses of 0.1–0.55 mg/m^2, with no maximum tolerated dose being reached. Five patients have demonstrated stable disease lasting for at least 2 months, including patients with melanoma, mantle cell lymphoma, Hodgkin's lymphoma, follicular lymphoma, and sarcoma [118].

Carfilzomib has been investigated in two phase 1 studies in patients with relapsed/refractory hematologic malignancies using two different administration schedules. In the first study, patients received carfilzomib 1.2–20 mg/m^2 on days 1–5 of 14-day cycles [119]. A total of 29 patients were treated, including 14 at doses at or above the minimal effective dose of 11 mg/m^2. Carfilzomib doses of $\geq$15 mg/m^2 resulted in a maximum of 80% inhibition of proteasome activity in peripheral blood and mononuclear cells [119], which is similar to what has been reported with bortezomib. Among the six multiple myeloma, three mantle cell lymphoma, one Waldenström's macroglobulinemia, and four non-Hodgkin's lymphoma patients treated with carfilzomib $\geq$11 mg/m^2, five responses were reported, including a partial response and two minimal responses in patients with multiple myeloma, a minimal response in the patient with Waldenström's macroglobulinemia, and a macroscopic complete response in a patient with gastrointestinal mantle cell lymphoma [119]. In addition, the four patients with non-Hodgkin's lymphoma experienced stable disease. Responses were reported in patients who had relapsed following prior bortezomib therapy. In total, 25% of patients reported grade 3 and 8% reported grade 4 adverse events during treatment; these toxicities were primarily hematologic. Dose-limiting toxicities included reversible febrile neutropenia and grade 4 thrombocytopenia. However, the majority of patient withdrawals were associated with either disease progression or the inconvenience of attending clinic for 5 successive days every 2 weeks; therefore, an alternative schedule is being used in phase 2 investigations [119]. This alternative schedule was investigated in a second phase 1 study and comprised twice-weekly consecutive day-dosing of carfilzomib 1.2–27 mg/m^2 on days 1, 2, 8, 9, 15, and 16 of 28-day cycles [120]. A total of 37 patients were treated in this study, including 16 at or above the minimal effective dose on this schedule of 15 mg/m^2. Five responses were seen in these 16 patients (13 with multiple myeloma and 3 with mantle cell lymphoma), including four partial and one minimal response in patients with multiple myeloma [120]. Six other patients, with multiple myeloma, T or B cell non-Hodgkin's lymphoma, or mantle cell lymphoma, had stable disease. As in the other phase 1 study of carfilzomib, responses were reported in patients who had previously been treated with bortezomib [120]. Such clinical activity with carfilzomib is perhaps not surprising in the context of the findings from various studies of bortezomib retreatment in patients with multiple myeloma, which have shown that a notable proportion of patients retain sensitivity and respond to bortezomib retreatment having relapsed following a response to initial bortezomib therapy [121–123].

Carfilzomib has also been investigated in a phase 1 study in patients with advanced solid tumors [124]. Patients received carfilzomib 20, 27, or 36 mg/m^2 on days 1, 2, 8, 9, 15, and 16 of 28-day cycles, following initial dosing on days 1 and 2 of cycle 1 of 20 mg/m^2. Among 14 patients treated to date, one patient with renal cancer achieved a partial response, and two patients, one with small cell lung cancer and one with mesothelioma, achieved stable disease lasting for more than 3 months [124].

Initial results have recently been reported from two ongoing phase 2 studies of carfilzomib in patients with relapsed [125] or relapsed and refractory multiple myeloma [126], in which patients received carfilzomib 20 mg/m^2 on days 1, 2, 8, 9, 15, and 16 of 4-week cycles, for up to 12 cycles. In the former study, 31 patients have been treated to date; 8 of 14 (57%) bortezomib-naïve patients have responded, including one complete, two very good partial, and five partial responses, and after a median follow-up of 3.6 months, none of these responders has progressed [125]. By contrast, only 3 of 17 (18%) patients who had previously been treated with bortezomib achieved a partial response, with one further patient achieving a minimal response; the median time to progression was not reached among these 17 patients after a median follow-up of 3.7 months [125]. In the other phase 2 study, 46 patients with relapsed and refractory multiple myeloma, all of whom had prior bortezomib treatment, have been treated to date. Among 39 evaluable patients, 5 (13%) have achieved a partial response and 5 (13%) a minimal response; these responders include five bortezomib-refractory patients [126].

No painful peripheral neuropathy was reported with carfilzomib in any of the phase 1 studies [119, 120, 124], and peripheral neuropathy was reportedly uncommon [125] and exacerbation of baseline neuropathy rare [126] in the two phase 2 studies of carfilzomib in multiple myeloma, with only one grade 3 peripheral neuropathy event reported [126]. Peripheral neuropathy is an important toxicity associated with bortezomib treatment [65, 66] that has been suggested as a class effect of proteasome inhibition [69]. However, given the established cumulative, dose-related, generally reversible nature of bortezomib-associated peripheral neuropathy [65, 66], it may be that the numbers of patients exposed and the cumulative doses of carfilzomib in the phase 1 studies and to date in the phase 2 studies were not sufficient to reveal this toxicity; larger studies at therapeutic doses, such as in expanded populations in the ongoing phase 2 studies, will be required to determine a complete safety profile for carfilzomib.

The second-generation proteasome inhibitors are being investigated in a number of other ongoing clinical studies that are yet to report data; the following were listed on ClinicalTrials.gov at the time of writing. A phase 1 study (NCT00572637) is underway investigating CEP-18770 in patients with solid tumors or non-Hodgkin's lymphoma, using the same dosing schedule as employed for bortezomib. Various studies of carfilzomib are in progress, including a phase 1 safety and pharmacokinetic study in patients with hematologic malignancies that is enrolling expansion cohorts for treatment with carfilzomib alone or in combination with dexamethasone (NCT00150462), phase 2 studies in relapsed (NCT00530816) and relapsed and refractory (NCT00511238) multiple myeloma, a phase 1 study in relapsed and refractory multiple myeloma patients with varying degrees of renal function (NCT00721734), a phase 1b study in combination with lenalidomide and dexamethasone in relapsed multiple myeloma (NCT00603447), and a phase 1b/2 study in relapsed solid tumors (NCT00531284). NPI-0052 is being investigated in an ongoing phase 1 study in combination with the histone deacetylase inhibitor vorinostat in patients with non-small cell lung cancer, pancreatic cancer, or melanoma (NCT00667082).

Notably, in all these clinical studies, the second-generation proteasome inhibitors are being administered by the intravenous route, although CEP-18770 and NPI-0052 have been shown to be orally active [71, 72, 91]. An oral analog of carfilzomib, PR-047 [82], is in development at Proteolix, although no clinical data have yet been published on this agent. Similarly, a second-generation proteasome inhibitor with the potential for oral administration is in development at Millennium Pharmaceuticals, Inc. It would appear that intravenous administration will remain the standard route for delivering proteasome inhibitor therapy for some time to come.

3 Other Therapeutic Targets in the Ubiquitin–Proteasome System

3.1 The Ubiquitin–Proteasome System Upstream of the 20S Proteasome

Protein degradation represents the culmination of several key steps that occur in the ubiquitin–proteasome system, including the recognition and preparation of proteins for degradation, and the interaction of these proteins with the 19S cap of the 26S proteasome. As these steps involve a variety of different enzymes, the upstream portion of the ubiquitin–proteasome system likely contains a number of druggable targets that could offer greater specificity, potentially boosting drug effectiveness and reducing nonspecific side effects, compared with 20S proteasome inhibition [127].

Figure 4 provides an overview of the cascade of events in the ubiquitin–proteasome system upstream of the 20S proteasome [127]. Among the key steps in protein degradation are the following [4–6, 15, 128, 129]: activation of ubiquitin by a ubiquitin-activating enzyme (E1); transfer of ubiquitin to a ubiquitin-conjugating enzyme (E2); recognition of a substrate protein for degradation by a ubiquitin–protein ligase (E3) and formation of a ligase–protein complex; polyubiquitination to mark the substrate for destruction by the 26S proteasome; recognition of the polyubiquitinated substrate by the 19S cap; deubiquitination; and catalytic destruction of the substrate by the 20S core.

Each of these events in the ubiquitin–proteasome system may represent a potential therapeutic target, which could have greater or lesser specificity depending on the extent of its involvement in the ubiquitin–proteasome system. For example, therapeutic approaches could include [127] inhibition of ubiquitin activation (E1 inhibition), inhibition of ubiquitin transfer to E2; inhibition of specific ligases (E3s) [130], blocking the recognition of polyubiquitinated proteins by the 19S cap [131], and inhibition of deubiquitinating enzymes [132, 133].

A number of other protein modifiers exist in addition to ubiquitin, including Nedd8, SUMO, and ISG15, each of which is activated by an E1 activating enzyme [129]. These proteins have specific downstream functions of relevance in different cancers, and therefore represent further possible therapeutic targets in the ubiquitin–proteasome system. Nedd8 is a ubiquitin-like protein that is activated by Nedd-8 activating enzyme (NAE). NAE transfers Nedd8 to the E2-conjugating enzyme UBC12, which subsequently transfers Nedd8 to the cullin subunit of a family of E3 ligases known as the Cullin-dependent ligases (CDLs) [127, 134–137]. The CDLs are responsible for recognizing at least hundreds of protein substrates marked for degradation [138], including various proteins with key roles in cell-cycle progression and signal transduction [139–141]. Inhibition of the activities of the CDLs impacts cellular processes of relevance to tumor cell growth and survival [139], and therefore represents a potential anticancer therapeutic strategy within the ubiquitin–proteasome system [140] that is more specific than 20S proteasome inhibition. This approach is being investigated by Millennium Pharmaceuticals, Inc., which is currently developing a first-in-class NAE inhibitor.

3.2 MLN4924: Nedd8-Activating Enzyme Inhibitor

MLN4924 is a small molecule inhibitor of NAE [139–143]; at the time of writing in December 2008, the structure was yet to be disclosed. Through inhibition of NAE, MLN4924 specifically inhibits the formation of Nedd8–Cullin complexes [140],

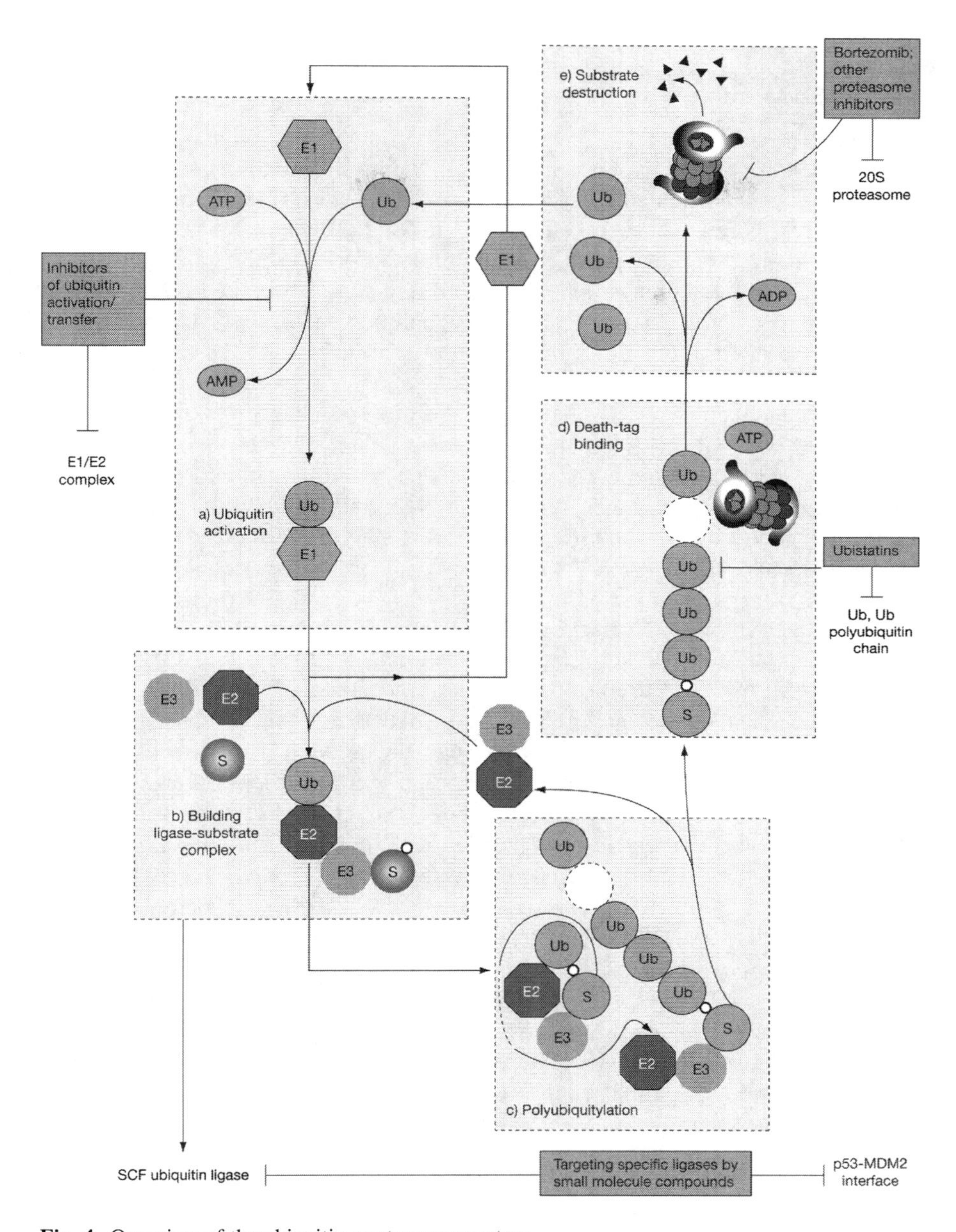

Fig. 4 Overview of the ubiquitin–proteasome system

blocking CDL activity and thereby resulting in the stabilization of CDL substrate proteins by preventing their ubiquitination and degradation [139, 140]. Among these substrate proteins are phosphorylated IκB (pIκB), the inhibitor of NF-κB, the DNA replication licensing factor Cdt1, Nrf-2, and cyclin D1 [139–141].

MLN4924 has been shown to have broad in vitro potency against several myeloma, lymphoma, and leukemia cell lines, including RPMI-8226, NCI-H929, WSU-DLCL2, Ly10, Ly19, HL-60, and MOLT-4, as well as multiple non-hematologic cell lines [142]. Preclinical studies have demonstrated that MLN4924 is a potent inhibitor of NF-κB signaling in vitro and in vivo through inhibition of CDL-mediated degradation of pIκB [139]. In NF-κB signaling-dependent OCI-Ly10 diffuse large B cell lymphoma (DLBCL) cells, MLN4924 resulted in marked stabilization of pIκBα, thus inhibiting NF-κB signaling and causing G1 arrest and acute induction of apoptosis [139]. By contrast, in OCI-Ly19 DLBCL cells, which are not dependent on NF-κB signaling for their survival, exposure to MLN4924 resulted in the elevation of multiple substrates of CDLs and the accumulation of S-phase cells with increased DNA content, followed by a DNA damage response and induction of cell death [139]. A similar mechanism of action has also been seen in other NF-κB signaling-independent tumor cell lines. In OCI-Ly10 and OCI-Ly19 mouse xenograft models, repeat in vivo administration of MLN4924 resulted in a pharmacodynamic response to NAE pathway inhibition, along with significant tumor growth inhibition; notably, in the OCI-Ly10 model, MLN4924 produced an elevation of pIκBα levels and induction of apoptosis, and tumor regressions were seen [139, 143].

The induction of a DNA damage response by MLN4924 has been further explored in the human colon carcinoma cell line HCT-116. In HCT-116 cells exposed to MLN4924, Cdt1 was stabilized and localized in the nucleus, resulting in an increase in DNA content through Cdt1-mediated over-replication, activation of a DNA damage checkpoint response through the ATM/ATR pathway and, ultimately, apoptosis [140, 142]. Cdt1 stabilization was also seen following in vivo administration of MLN4924 in mouse xenograft models of lymphoma and lung, breast, and colon carcinoma [141]. Tumor growth inhibition was seen in the HCT-116 colon carcinoma model, associated with elevated levels of Nrf-2, cyclin D1, and Cdt1 and evidence of ATM/ATR activation and a DNA damage response. Increases in cleaved caspase-3 levels and the number of apoptotic cells were also seen. These pharmacodynamic markers are now being adapted to support clinical development of MLN4924 [141]. Two phase 1 trials are underway; in a trial in patients with lymphoma or MM (NCT00722488), MLN4924 is being administered intravenously using three different dosing schedules in 21-day cycles, while in a trial in patients with nonhematologic malignancies (NCT00677170), a schedule comprising intravenous administration for 5 consecutive days in 21-day cycles is being employed.

4 Conclusion

Over the past decade, proteasome inhibition has emerged as a highly successful therapeutic approach, despite initial doubts that the proteasome could represent a druggable target. Bortezomib has proven clinically successful for the treatment of multiple myeloma and other hematologic malignancies; consequently, the number of drugs in development that target components of the ubiquitin–proteasome system has expanded markedly to include second-generation proteasome inhibitors and a newer class of molecule that inhibits a novel component of the ubiquitin–proteasome system. Numerous companies are now working in this area with the aim of improving efficacy and outcomes while reducing the adverse effects of therapy and improving the convenience of administration. As our understanding of the complexities of the ubiquitin–proteasome system continues to improve, further targets, offering increasing specificity, are likely to emerge. Ultimately, these developments and subsequent generations of as-yet-unknown novel agents are expected to improve outcomes for patients with cancer.

5 Notes Added in Proof - Novel Second-Generation Proteasome Inhibitors

Since the initial development of this chapter (Feb 2009), preclinical and clinical data have been reported for the additional, novel second-generation proteasome inhibitors MLN9708 [144] and ONX0912 [145, 146]. MLN9708 is a small molecule inhibitor of the proteasome that hydrolyzes immediately in aqueous solution or plasma to MLN2238, the biologically active form [144]. MLN2238, an N-capped dipeptidyl leucine boronic acid, primarily inhibits the β5 (chymotrypsin-like) proteolytic site of the proteasome with an IC_{50} of 3.4 nmol/L; the β1 (caspase-like) and β2 (trypsin-like) sites are inhibited only at higher concentrations (IC_{50} values of 31 and 3,500 nmol/L, respectively) [144]. Like bortezomib, MLN2238 is a reversible inhibitor of the proteasome; however, the dissociation half-life of MLN2238 is approximately 6-fold shorter than that of bortezomib, at 18 vs. 110 min, and thus proteasome inhibition with MLN2238 is much more readily reversible [144]. In tumor cell line studies, MLN2238 strongly inhibited TNFα-induced NF-κB activity and showed antiproliferative activity in a number of tumor cell types [144]. Reflecting the differences in binding kinetics between MLN2238 and bortezomib, in vivo studies in mice and rats showed that MLN2238 had larger blood volume distribution at steady state and improved distribution into the tissue compartment; consequently, mouse xenograft models showed greater maximum and sustained tumor proteasome inhibition with MLN2238 vs. bortezomib [144]. In addition, MLN2238 demonstrated oral bioavailability [144]. MLN2238 showed similar antitumor activity to bortezomib in a CWR22 (human prostate) model and greater activity in WSU-DLCL2 and OCI-Ly7-Luc lymphoma models, including significantly prolonged survival vs. vehicle-treated controls in the latter model [144].

ONX0912, a peptide epoxyketone, is an irreversible inhibitor of the 20S proteasome, primarily the β5 proteolytic site (IC_{50} of 36 nM) [145]. It also inhibits the $\beta5_i$ site of the immunoproteasome with an IC_{50} of 82 nM [145]. ONX0912 demonstrated moderate absolute oral bioavailability across multiple species, and showed rapid absorption and distribution following oral administration in animals, with proteasome inhibition of more than 80% in blood and tissue within 15 min of dosing [145]. Cytotoxicity has been reported in Waldenström's macroglobulinemia cells (BCWM.1) and low-grade lymphoma cell lines (RL, MEC.1) associated with the inhibition of the chymotrypsin-like activity of both the proteasome and immunoproteasome [146], while in in vivo studies, ONX0912 showed similar antitumor activity to carfilzomib in mouse xenograft models of non-Hodgkin's lymphoma (RL) and colorectal cancer (CT-26) [145].

Further to the immunoproteasome inhibitory activity of ONX0912, initial preclinical data have also been reported for two novel specific inhibitors of the immunoproteasome, PR-957 [147] and IPSI-001 [148]. Like ONX0912, PR-957 is a peptide epoxyketone that irreversibly inhibits the $\beta5_i$ site of the immunoproteasome with 20- to 40-fold greater selectivity compared with the $\beta5$ and $\beta1_i$ sites of the proteasome and immunoproteasome, respectively [147]. IPSI-001 is a peptidyl aldehyde compound that potently inhibits the $\beta5_i$ (K_i 1.03 μM) and $\beta1_i$ (K_i 1.45 μM) sites of the immunoproteasome while showing lower potency for the $\beta5$ (K_i 105 μM) and $\beta1$ (K_i 239 μM) sites of the proteasome [148].

MLN9708 and ONX0912 have entered clinical trials, and data have been reported from the first-in-human phase 1 study of MLN9708 in advanced nonhematologic malignancies [149]. In the dose-escalation component of this study, 23 patients with colon, skin, renal cancer, sarcoma, and other malignancies received MLN9708 at 0.125–2.34 mg/m^2 on days 1, 4, 8, and 11 of 21-day cycles; the maximum tolerated dose was determined to be 1.76 mg/m^2, and MLN9708 was well tolerated up to this dose on this schedule, with a safety profile reflecting some aspects of that seen with bortezomib [149]. Pharmacokinetic and pharmacodynamic studies showed rapid disposition of MLN9708 and sustained proteasome inhibition in blood following multiple dosing [149]. No formal efficacy data have been reported to date; patients with non-small-cell lung cancer, head and neck cancer, soft tissue sarcoma, and prostate cancer are being enrolled to expanded cohorts at the maximum tolerated dose [149]. In other phase 1 studies, MLN9708 is being investigated via intravenous administration in patients with relapsed and/or refractory lymphoma (NCT00893464) and via oral administration in patients with relapsed and refractory multiple myeloma (NCT00932698 and NCT00963820). ONX0912 is being investigated via oral administration in patients with advanced refractory or recurrent solid tumors (NCT01129349).

References

1. Adams J (2004) The proteasome: a suitable antineoplastic target. Nat Rev Cancer 4:349–360
2. Adams J (2004) The development of proteasome inhibitors as anticancer drugs. Cancer Cell 5:417–421
3. Adams J, Kauffman M (2004) Development of the proteasome inhibitor Velcade (Bortezomib). Cancer Invest 22:304–311

4. Ciechanover A (1998) The ubiquitin-proteasome pathway: on protein death and cell life. EMBO J 17:7151–7160
5. Ciechanover A, Schwartz AL (1998) The ubiquitin-proteasome pathway: the complexity and myriad functions of proteins death. Proc Natl Acad Sci U S A 95:2727–2730
6. Hershko A (2005) The ubiquitin system for protein degradation and some of its roles in the control of the cell division cycle. Cell Death Differ 12:1191–1197
7. Goldberg AL, Elledge SJ, Harper JW (2001) The cellular chamber of doom. Sci Am 284:68–73
8. Nencioni A, Grunebach F, Patrone F, Ballestrero A, Brossart P (2007) Proteasome inhibitors: antitumor effects and beyond. Leukemia 21:30–36
9. Myung J, Kim KB, Crews CM (2001) The ubiquitin-proteasome pathway and proteasome inhibitors. Med Res Rev 21:245–273
10. Kisselev AF, Goldberg AL (2001) Proteasome inhibitors: from research tools to drug candidates. Chem Biol 8:739–758
11. Adams J, Behnke M, Chen S et al (1998) Potent and selective inhibitors of the proteasome: dipeptidyl boronic acids. Bioorg Med Chem Lett 8:333–338
12. Adams J, Palombella VJ, Sausville EA et al (1999) Proteasome inhibitors: a novel class of potent and effective antitumor agents. Cancer Res 59:2615–2622
13. Millennium Pharmaceuticals Inc. (2009) VELCADE® (bortezomib) for Injection. Prescribing information. Cambridge, MA. Issued December 2009, Rev 10
14. Wilk S, Orlowski M (1983) Evidence that pituitary cation-sensitive neutral endopeptidase is a multicatalytic protease complex. J Neurochem 40:842–849
15. Adams J (2003) The proteasome: structure, function, and role in the cell. Cancer Treat Rev 29(Suppl 1):3–9
16. Groll M, Berkers CR, Ploegh HL, Ovaa H (2006) Crystal structure of the boronic acid-based proteasome inhibitor bortezomib in complex with the yeast 20S proteasome. Structure 14:451–456
17. Richardson PG, Mitsiades C, Schlossman R et al (2008) Bortezomib in the front-line treatment of multiple myeloma. Expert Rev Anticancer Ther 8:1053–1072
18. Obeng EA, Carlson LM, Gutman DM et al (2006) Proteasome inhibitors induce a terminal unfolded protein response in multiple myeloma cells. Blood 107:4907–4916
19. Baumann P, Mandl-Weber S, Oduncu F, Schmidmaier R (2008) Alkylating agents induce activation of NFkappaB in multiple myeloma cells. Leuk Res 32:1144–1147
20. Hideshima T, Richardson P, Chauhan D et al (2001) The proteasome inhibitor PS-341 inhibits growth, induces apoptosis, and overcomes drug resistance in human multiple myeloma cells. Cancer Res 61:3071–3076
21. Ma MH, Yang HH, Parker K et al (2003) The proteasome inhibitor PS-341 markedly enhances sensitivity of multiple myeloma tumor cells to chemotherapeutic agents. Clin Cancer Res 9:1136–1144
22. Mitsiades N, Mitsiades CS, Richardson PG et al (2003) The proteasome inhibitor PS-341 potentiates sensitivity of multiple myeloma cells to conventional chemotherapeutic agents: therapeutic applications. Blood 101:2377–2380
23. Catley L, Weisberg E, Kiziltepe T et al (2006) Aggresome induction by proteasome inhibitor bortezomib and alpha-tubulin hyperacetylation by tubulin deacetylase (TDAC) inhibitor LBH589 are synergistic in myeloma cells. Blood 108:3441–3449
24. Pei XY, Dai Y, Grant S (2004) Synergistic induction of oxidative injury and apoptosis in human multiple myeloma cells by the proteasome inhibitor bortezomib and histone deacetylase inhibitors. Clin Cancer Res 10:3839–3852
25. Chauhan D, Velankar M, Brahmandam M et al (2007) A novel Bcl-2/Bcl-X(L)/Bcl-w inhibitor ABT-737 as therapy in multiple myeloma. Oncogene 26:2374–2380
26. Gomez-Bougie P, Maiga S, Pellat-Deceunynck C et al (2006) The pan-Bcl-2 inhibitor GX15-O70 induces apoptosis in human myeloma cells by Noxa induction and strongly enhances melphalan, bortezomib or TRAIL-R1 antibody apoptotic effect [abstract]. Blood 108:991a

27. Perez-Galan P, Roue G, Villamor N, Campo E, Colomer D (2007) The BH3-mimetic GX15-070 synergizes with bortezomib in mantle cell lymphoma by enhancing Noxa-mediated activation of Bak. Blood 109:4441–4449
28. Liu FT, Agrawal SG, Gribben JG et al (2008) Bortezomib blocks Bax degradation in malignant B cells during treatment with TRAIL. Blood 111:2797–2805
29. Nencioni A, Wille L, Dal BG et al (2005) Cooperative cytotoxicity of proteasome inhibitors and tumor necrosis factor-related apoptosis-inducing ligand in chemoresistant Bcl-2-overexpressing cells. Clin Cancer Res 11:4259–4265
30. Roue G, Perez-Galan P, Lopez-Guerra M et al (2007) Selective inhibition of IkappaB kinase sensitizes mantle cell lymphoma B cells to TRAIL by decreasing cellular FLIP level. J Immunol 178:1923–1930
31. San Miguel JF, Schlag R, Khuageva NK et al (2008) Bortezomib plus melphalan and prednisone for initial treatment of multiple myeloma. N Engl J Med 359:906–917
32. Richardson PG, Mitsiades C, Ghobrial I, Anderson K (2006) Beyond single-agent bortezomib: combination regimens in relapsed multiple myeloma. Curr Opin Oncol 18:598–608
33. Richardson PG, Hideshima T, Mitsiades C, Anderson KC (2007) The emerging role of novel therapies for the treatment of relapsed myeloma. J Natl Compr Canc Netw 5:149–162
34. Kahl BS, Peterson C, Blank J et al (2007) A feasibility study of VcR-CVAD with maintenance rituximab for untreated mantle cell lymphoma [abstract]. J Clin Oncol 25:456s
35. Mounier N, Ribrag V, Haioun C et al (2007) Efficacy and toxicity of two schedules of R-CHOP plus bortezomib in front-line B lymphoma patients: a randomized phase II trial from the Groupe d'Etude des Lymphomes de l'Adulte (GELA) [abstract]. J Clin Oncol 25:8010
36. Belch A, Kouroukis CT, Crump M et al (2007) A phase II study of bortezomib in mantle cell lymphoma: the National Cancer Institute of Canada Clinical Trials Group trial IND.150. Ann Oncol 18:116–121
37. Drach J, Kaufmann H, Pichelmayer O et al (2007) Bortezomib, rituximab, and dexamethasone (BORID) as salvage treatment in relapsed/refractory mantle cell lymphoma: sustained disease control in patients achieving a complete remission [abstract]. Blood 110:760a
38. Fisher RI, Bernstein SH, Kahl BS et al (2006) Multicenter phase II study of bortezomib in patients with relapsed or refractory mantle cell lymphoma. J Clin Oncol 24:4867–4874
39. Goy A, Bernstein S, Kahl B et al (2007) Durable responses with bortezomib in patients with relapsed or refractory mantle cell lymphoma (MCL): updated time-to-event analyses of the multicenter PINNACLE study [abstract]. Blood 110:45a
40. O'Connor OA, Wright J, Moskowitz C et al (2005) Targeting the proteasome pathway with bortezomib in patients with mantle cell (MCL) and follicular lymphoma (FL) produces prolonged progression free survival among responding patients: results of a multicenter phase II experience [abstract]. Ann Oncol 16(Suppl 5):v66
41. Goy A, Younes A, McLaughlin P et al (2005) Phase II study of proteasome inhibitor bortezomib in relapsed or refractory B-cell non-Hodgkin's lymphoma. J Clin Oncol 23:667–675
42. O'Connor OA, Wright J, Moskowitz C et al (2005) Phase II clinical experience with the novel proteasome inhibitor bortezomib in patients with indolent non-Hodgkin's lymphoma and mantle cell lymphoma. J Clin Oncol 23:676–684
43. Strauss SJ, Maharaj L, Hoare S et al (2006) Bortezomib therapy in patients with relapsed or refractory lymphoma: potential correlation of in vitro sensitivity and tumor necrosis factor alpha response with clinical activity. J Clin Oncol 24:2105–2112
44. De Vos S, Dakhil SR, McLaughlin P et al (2006) Phase 2 study of bortezomib weekly or twice weekly plus rituximab in patients with follicular (FL) or marginal zone (MZL) lymphoma: final results [abstract]. Blood 108:208a
45. Chen CI, Kouroukis CT, White D et al (2007) Bortezomib is active in patients with untreated or relapsed Waldenstrom's macroglobulinemia: a phase II study of the National Cancer Institute of Canada Clinical Trials Group. J Clin Oncol 25:1570–1575

46. Dimopoulos MA, Anagnostopoulos A, Kyrtsonis MC et al (2005) Treatment of relapsed or refractory Waldenstrom's macroglobulinemia with bortezomib. Haematologica 90:1655–1658
47. Treon SP, Hunter ZR, Matous J et al (2007) Multicenter clinical trial of bortezomib in relapsed/refractory Waldenstrom's macroglobulinemia: results of WMCTG Trial 03-248. Clin Cancer Res 13:3320–3325
48. Treon SP, Ioakimidis L, Soumerai JD et al (2008) Primary therapy of Waldenstrom's macroglobulinemia with bortezomib, dexamethasone and rituximab: results of WMCTG clinical trial 05-180 [abstract]. J Clin Oncol 26:8519
49. Kastritis E, Anagnostopoulos A, Roussou M et al (2007) Treatment of light chain (AL) amyloidosis with the combination of bortezomib and dexamethasone. Haematologica 92:1351–1358
50. Wechalekar AD, Lachmann HJ, Offer M, Hawkins PN, Gillmore JD (2008) Efficacy of bortezomib in systemic AL amyloidosis with relapsed/refractory clonal disease. Haematologica 93:295–298
51. Chauhan D, Li G, Shringarpure R et al (2003) Blockade of Hsp27 overcomes bortezomib/proteasome inhibitor PS-341 resistance in lymphoma cells. Cancer Res 63:6174–6177
52. Hideshima T, Chauhan D, Ishitsuka K et al (2005) Molecular characterization of PS-341 (bortezomib) resistance: implications for overcoming resistance using lysophosphatidic acid acyltransferase (LPAAT)-beta inhibitors. Oncogene 24:3121–3129
53. McConkey DJ, Zhu K (2008) Mechanisms of proteasome inhibitor action and resistance in cancer. Drug Resist Updat 11:164–179
54. Yang DT, Young KH, Kahl BS, Miyoshi S (2007) Bortezomib resistant constitutive NF-κB activation in mantle cell lymphoma [abstract]. Blood 110:1015a
55. Dreicer R, Petrylak D, Agus D, Webb I, Roth B (2007) Phase I/II study of bortezomib plus docetaxel in patients with advanced androgen-independent prostate cancer. Clin Cancer Res 13:1208–1215
56. Hainsworth JD, Meluch AA, Spigel DR et al (2007) Weekly docetaxel and bortezomib as first-line treatment for patients with hormone-refractory prostate cancer: a Minnie Pearl Cancer Research Network phase II trial. Clin Genitourin Cancer 5:278–283
57. Fanucchi MP, Fossella FV, Belt R et al (2006) Randomized phase II study of bortezomib alone and bortezomib in combination with docetaxel in previously treated advanced non-small-cell lung cancer. J Clin Oncol 24:5025–5033
58. Lara PN Jr, Koczywas M, Quinn DI et al (2006) Bortezomib plus docetaxel in advanced non-small cell lung cancer and other solid tumors: a phase I California Cancer Consortium trial. J Thorac Oncol 1:126–134
59. Lynch TJ, Fenton DW, Hirsh V et al (2007) Randomized phase II study of erlotinib alone and in combination with bortezomib in previously treated advanced non-small cell lung cancer (NSCLC) [abstract]. J Clin Oncol 25:429s
60. Ikezoe T, Yang Y, Saito T, Koeffler HP, Taguchi H (2004) Proteasome inhibitor PS-341 down-regulates prostate-specific antigen (PSA) and induces growth arrest and apoptosis of androgen-dependent human prostate cancer LNCaP cells. Cancer Sci 95:271–275
61. Williams S, Pettaway C, Song R et al (2003) Differential effects of the proteasome inhibitor bortezomib on apoptosis and angiogenesis in human prostate tumor xenografts. Mol Cancer Ther 2:835–843
62. Scagliotti G (2006) Proteasome inhibitors in lung cancer. Crit Rev Oncol Hematol 58: 177–189
63. Schenkein DP (2005) Preclinical data with bortezomib in lung cancer. Clin Lung Cancer 7 (Suppl 2):S49–S55
64. Voortman J, Checinska A, Giaccone G (2007) The proteasomal and apoptotic phenotype determine bortezomib sensitivity of non-small cell lung cancer cells. Mol Cancer 6:73
65. Richardson PG, Briemberg H, Jagannath S et al (2006) Frequency, characteristics, and reversibility of peripheral neuropathy during treatment of advanced multiple myeloma with bortezomib. J Clin Oncol 24:3113–3120

66. San Miguel JF, Richardson P, Sonneveld P et al (2005) Frequency, characteristics, and reversibility of peripheral neuropathy (PN) in the APEX trial [abstract]. Blood 106:111a
67. Csizmadia V, Raczynski A, Csizmadia E et al (2008) Effect of an experimental proteasome inhibitor on the cytoskeleton, cytosolic protein turnover, and induction in the neuronal cells in vitro. Neurotoxicology 29:232–243
68. Silverman L, Csizmadia V, Kadambi VJ et al (2006) Model for proteasome inhibition associated peripheral neuropathy [abstract]. Toxicol Pathol 34:989
69. Silverman L, Csizmadia V, Brewer K, Simpson C, Alden C (2008) Proteasome inhibitor associated neuropathy is mechanism based [abstract]. Blood 112: Abstract 2646.
70. Moreau P, Coiteux V, Hulin C et al (2008) Prospective comparison of subcutaneous versus intravenous administration of bortezomib in patients with multiple myeloma. Haematologica 93:1908–1911
71. Dorsey BD, Iqbal M, Chatterjee S et al (2008) Discovery of a potent, selective, and orally active proteasome inhibitor for the treatment of cancer. J Med Chem 51:1068–1072
72. Piva R, Ruggeri B, Williams M et al (2008) CEP-18770: a novel, orally active proteasome inhibitor with a tumor-selective pharmacologic profile competitive with bortezomib. Blood 111:2765–2775
73. Demo SD, Kirk CJ, Aujay MA et al (2007) Antitumor activity of PR-171, a novel irreversible inhibitor of the proteasome. Cancer Res 67:6383–6391
74. Feling RH, Buchanan GO, Mincer TJ et al (2003) Salinosporamide A: a highly cytotoxic proteasome inhibitor from a novel microbial source, a marine bacterium of the new genus salinospora. Angew Chem Int Ed Engl 42:355–357
75. Tsueng G, Teisan S, Lam KS (2008) Defined salt formulations for the growth of Salinispora tropica strain NPS21184 and the production of salinosporamide A (NPI-0052) and related analogs. Appl Microbiol Biotechnol 78:827–832
76. Groll M, Huber R, Potts BC (2006) Crystal structures of salinosporamide A (NPI-0052) and B (NPI-0047) in complex with the 20S proteasome reveal important consequences of beta-lactone ring opening and a mechanism for irreversible binding. J Am Chem Soc 128:5136–5141
77. Lightcap ES, McCormack TA, Pien CS et al (2000) Proteasome inhibition measurements: clinical application. Clin Chem 46:673–683
78. Miller CP, Ban K, Dujka ME et al (2007) NPI-0052, a novel proteasome inhibitor, induces caspase-8 and ROS-dependent apoptosis alone and in combination with HDAC inhibitors in leukemia cells. Blood 110:267–277
79. Ruiz S, Krupnik Y, Keating M et al (2006) The proteasome inhibitor NPI-0052 is a more effective inducer of apoptosis than bortezomib in lymphocytes from patients with chronic lymphocytic leukemia. Mol Cancer Ther 5:1836–1843
80. Roccaro AM, Leleu X, Sacco A et al (2008) Dual targeting of the proteasome regulates survival and homing in Waldenstrom's macroglobulinemia. Blood 111:4752–4763
81. Singh AV, Lloyd GK, Palladino MA, Chauhan D, Anderson KC (2008) Pharmacodynamic and efficacy studies of a novel proteasome inhibitor NPI-0052 in human plasmacytoma xenograft mouse model [abstract]. Blood 112: Abstract 3665
82. Muchamuel T, Aujay M, Bennett MK et al (2008) Preclinical pharmacology and in vitro characterization of PR-047, an oral inhibitor of the 20S proteasome [abstract]. Blood 112: Abstract 3671
83. Williamson MJ, Blank JL, Bruzzese FJ et al (2006) Comparison of biochemical and biological effects of ML858 (salinosporamide A) and bortezomib. Mol Cancer Ther 5: 3052–3061
84. Kisselev AF, Callard A, Goldberg AL (2006) Importance of the different proteolytic sites of the proteasome and the efficacy of inhibitors varies with the protein substrate. J Biol Chem 281:8582–8590
85. Arastu-Kapur S, Shenk K, Parlati F, Bennett MK (2008) Non-proteasomal targets of proteasome inhibitors bortezomib and carfilzomib [abstract]. Blood 112: Abstract 2657

86. Hideshima T, Mitsiades C, Akiyama M et al (2003) Molecular mechanisms mediating antimyeloma activity of proteasome inhibitor PS-341. Blood 101:1530–1534
87. Kukreja A, Hutchinson A, Mazumder A et al (2007) Bortezomib disrupts tumour-dendritic cell interactions in myeloma and lymphoma: therapeutic implications. Br J Haematol 136:106–110
88. Mitsiades N, Mitsiades CS, Poulaki V et al (2002) Molecular sequelae of proteasome inhibition in human multiple myeloma cells. Proc Natl Acad Sci U S A 99:14374–14379
89. Strauss SJ, Higginbottom K, Juliger S et al (2007) The proteasome inhibitor bortezomib acts independently of p53 and induces cell death via apoptosis and mitotic catastrophe in B-cell lymphoma cell lines. Cancer Res 67:2783–2790
90. Landowski TH, Megli CJ, Nullmeyer KD, Lynch RM, Dorr RT (2005) Mitochondrial-mediated disregulation of Ca2+ is a critical determinant of Velcade (PS-341/bortezomib) cytotoxicity in myeloma cell lines. Cancer Res 65:3828–3836
91. Chauhan D, Catley L, Li G et al (2005) A novel orally active proteasome inhibitor induces apoptosis in multiple myeloma cells with mechanisms distinct from bortezomib. Cancer Cell 8:407–419
92. Ahn KS, Sethi G, Chao TH et al (2007) Salinosporamide A (NPI-0052) potentiates apoptosis, suppresses osteoclastogenesis, and inhibits invasion through down-modulation of NF-kappaB regulated gene products. Blood 110:2286–2295
93. Barral AM, Chao T-H, Kanabolooki S et al (2007) The proteasome inhibitor NPI-0052 reduces tumor growth and overcomes resistance of prostate cancer to rhTRAIL via inhibition of the NF-kß pathway [abstract]. Proc AACR 2007, Annual Meeting: Abstract 1465
94. Cusack JC Jr, Liu R, Xia L et al (2006) NPI-0052 enhances tumoricidal response to conventional cancer therapy in a colon cancer model. Clin Cancer Res 12:6758–6764
95. Baritaki S, Suzuki E, Umezawa K et al (2008) Inhibition of Yin Yang 1-dependent repressor activity of DR5 transcription and expression by the novel proteasome inhibitor NPI-0052 contributes to its TRAIL-enhanced apoptosis in cancer cells. J Immunol 180:6199–6210
96. Baritaki S, Chapman A, Wu K et al (2008) The novel proteasome inhibitor NPI-0052 induces the expression of Raf-1 kinase inhibitor protein (RKIP) in B-NHL via inhibition of the transcription repressor Snail: roles of Snail and RKIP in sensitization to TRAIL apoptosis [abstract]. Blood 112: Abstract 2611
97. Kuhn DJ, Chen Q, Voorhees PM et al (2007) Potent activity of carfilzomib, a novel, irreversible inhibitor of the ubiquitin-proteasome pathway, against preclinical models of multiple myeloma. Blood 110:3281–3290
98. Chao T-H, Barral AM, Lloyd KG et al (2008) The pharmacodynamic profile of NPI-0052 is cell-type specific [abstract]. Proc AACR 2008, Annual Meeting: Abstract 3257
99. Stapnes C, Doskeland AP, Hatfield K et al (2007) The proteasome inhibitors bortezomib and PR-171 have antiproliferative and proapoptotic effects on primary human acute myeloid leukaemia cells. Br J Haematol 136:814–828
100. Stewart AK, Sullivan D, Lonial S et al (2006) Pharmacokinetic (PK) and pharmacodynamics (PD) study of two doses of bortezomib (Btz) in patients with relapsed multiple myeloma (MM) [abstract]. Blood 108:1008a
101. Yang J, Fonseca F, Ho M et al (2006) Metabolism, disposition and pharmacokinetics of PR-171, a novel inhibitor of the 20S proteasome [abstract]. Blood 108: Abstract 5067
102. Kirk CJ, Jiang J, Muchamuel T et al (2008) The selective proteasome inhibitor carfilzomib is well tolerated in experimental animals with dose intensive administration [abstract]. Blood 112: Abstract 2765
103. Buglio D, Georgakis G, Yazbeck V et al (2007) A novel proteasome inhibitor, NPI-0052 is active in Hodgkin and mantle cell lymphoma cell lines [abstract]. Proc AACR 2007, Annual Meeting: Abstract 1454
104. Suzuki E, Palladino M, Cheng G, Bonavida B (2007) Sensitization of B-NHL resistant tumor cells overexpressing Bcl-xL to TRAIL-induced apoptosis by the novel proteasome inhibitor

Salinosporamide A (NPI-0052) [abstract]. Proc AACR 2007, Annual Meeting: Abstract 1445
105. Sanchez E, Campbell RA, Steinberg JA et al (2008) The novel proteasome inhibitor CEP-18770 inhibits myeloma tumor growth *in vitro* and *in vivo* and enhances the anti-MM effects of melphalan [abstract]. Blood 112: Abstract 843
106. Huang X, Bailey K, Di Liberto M et al (2008) Induction of sustained early G1 arrest by selective inhibition of CDK4 and CDK6 primes myeloma cells for synergistic killing by proteasome inhibitors carfilzomib and PR-047 [abstract]. Blood 112: Abstract 3670
107. Paoluzzi L, Gonen M, Gardner JR et al (2008) Targeting Bcl-2 family members with the BH3 mimetic AT-101 markedly enhances the therapeutic effects of chemotherapeutic agents in in vitro and in vivo models of B-cell lymphoma. Blood 111:5350–5358
108. Dajee M, Aujay M, Demo S et al (2008) The selective proteasome inhibitor carfilzomib in combination with chemotherapeutic agents improves anti-tumor response in solid tumor xenograft models [abstract]. Eur J Cancer 6:75
109. Chauhan D, Singh AV, Brahmandam M et al (2008) Combination of a novel proteasome inhibitor NPI-0052 and lenalidomide trigger in vivo synergistic cytotoxicity in multiple myeloma [abstract]. Blood 112: Abstract 3662
110. Miller CP, Palladino M, Chandra J (2008) Overlapping functional activities of proteasome inhibitor, NPI-0052, and HDAC inhibitors contribute to synergistic cytotoxicity in leukemia cells [abstract]. Proc AACR 2008, Annual Meeting: Abstract 3255
111. Yazbeck VY, Georgakis GV, Li Y et al (2006) Inhibition of the pan-Bcl-2 family by the small molecule GX15-070 induces apoptosis in mantle cell lymphoma (MCL) cells and enhances the activity of two proteasome inhibitors (NPI-0052 and bortezomib), and doxorubicin chemotherapy [abstract]. Blood 108:716a
112. Sundi D, Fournier KF, Wray CJ, Marquis LM, McConkey DJ (2008) Combination treatment of pancreatic cancer cells with NPI-0052 and TRAIL activates heterogeneous apoptotic responses [abstract]. Proc ASCO 2008; Gastrointestinal Cancers Symposium: Abstract 183.
113. Wray C, Fournier KF, Sundi D, Marquis LM, McConkey DJ (2008) Combination proteasome- and histone deacetylase (HDAC) inhibitor treatment of pancreatic cancer [abstract]. Proc ASCO 2008; Gastrointestinal Cancers Symposium: Abstract 259
114. Cusack JC, Liu R, Xia L, Neuteboom S, Palladino M (2008) Salinosporamide A, a novel orally active proteasome inhibitor NPI-0052 enhances tumoricidal response to multidrug chemo- and molecular therapy regimens in a pancreatic cancer xenograft model [abstract]. Proc ASCO 2008; Gastrointestinal Cancers Symposium: Abstract 93
115. Chauhan D, Singh A, Brahmandam M et al (2008) Combination of proteasome inhibitors bortezomib and NPI-0052 trigger in vivo synergistic cytotoxicity in multiple myeloma. Blood 111:1654–1664
116. Richardson P, Hofmeister CC, Zimmerman TM et al (2008) Phase 1 clinical trial of NPI-0052, a novel proteasome inhibitor in patients with multiple myeloma [abstract]. Blood 112: Abstract 2770
117. Kurzrock R, Hamlin P, Gordon M et al (2008) NPI-0052 (a 2nd generation proteasome inhibitor) phase 1 study in patients with lymphoma and solid tumors [abstract]. Eur J Cancer 6:73–74
118. Townsend A, Padrik P, Mainwaring P et al (2008) Phase I clinical trial of the 2nd generation proteasome inhibitor NPI-0052 in patients with advanced malignancies with a CLL RP2D cohort [abstract]. Eur J Cancer 6:74
119. Orlowski RZ, Stewart K, Vallone M et al (2007) Safety and antitumor efficacy of the proteasome inhibitor carfilzomib (PR-171) dosed for five consecutive days in hematologic malignancies: phase 1 results [abstract]. Blood 110:127a
120. Alsina M, Trudel S, Vallone M et al (2007) Phase 1 single agent antitumor activity of twice weekly consecutive day dosing of the proteasome inhibitor carfilzomib (PR-171) in hematologic malignancies [abstract]. Blood 110:128a

121. Conner TM, Doan QD, Walters IB, LeBlanc AL, Beveridge RA (2008) An observational, retrospective analysis of retreatment with bortezomib for multiple myeloma. Clin Lymphoma Myeloma 8:140–145
122. Sood R, Carloss H, Kerr R et al (2006) Retreatment with bortezomib alone or in combination for patients with multiple myeloma (MM) following an initial response to bortezomib: a phase IV, open-label trial [abstract]. Ann Oncol 17:ix205–ix206
123. Wolf J, Richardson PG, Schuster M et al (2008) Utility of bortezomib retreatment in relapsed or refractory multiple myeloma patients: a multicenter case series. Clin Adv Hematol Oncol 6:755–760
124. Papadopoulos KP, Infante JR, Wong AF et al (2008) A phase I safety, pharmacokinetic and pharmacodynamic study of carfilzomib, a selective proteasome inhibitor, in subjects with advanced solid tumors [abstract]. Eur J Cancer 6:121
125. Vij R, Wang M, Orlowski R et al (2008) Initial results of PX-171-004, an open-label, single-arm, phase II study of carfilzomib (CFZ) in patients with relapsed myeloma (MM) [abstract]. Blood 112: Abstract 865
126. Jagannath S, Vij R, Stewart AK et al (2008) Initial results of PX-171-003, an open-label, single-arm, phase II study of carfilzomib (CFZ) in patients with relapsed and refractory multiple myeloma (MM) [abstract]. Blood 112: Abstract 864
127. Nalepa G, Rolfe M, Harper JW (2006) Drug discovery in the ubiquitin-proteasome system. Nat Rev Drug Discov 5:596–613
128. Adams J (2002) Development of the proteasome inhibitor PS-341. Oncologist 7:9–16
129. Herrmann J, Lerman LO, Lerman A (2007) Ubiquitin and ubiquitin-like proteins in protein regulation. Circ Res 100:1276–1291
130. Sun Y (2003) Targeting E3 ubiquitin ligases for cancer therapy. Cancer Biol Ther 2:623–629
131. Verma R, Peters NR, D'Onofrio M et al (2004) Ubistatins inhibit proteasome-dependent degradation by binding the ubiquitin chain. Science 306:117–120
132. Komada M (2008) Controlling receptor downregulation by ubiquitination and deubiquitination. Curr Drug Discov Technol 5:78–84
133. Nicholson B, Marblestone JG, Butt TR, Mattern MR (2007) Deubiquitinating enzymes as novel anticancer targets. Future Oncol 3:191–199
134. Chiba T, Tanaka K (2004) Cullin-based ubiquitin ligase and its control by NEDD8-conjugating system. Curr Protein Pept Sci 5:177–184
135. Pan ZQ, Kentsis A, Dias DC, Yamoah K, Wu K (2004) Nedd8 on cullin: building an expressway to protein destruction. Oncogene 23:1985–1997
136. Petroski MD, Deshaies RJ (2005) Function and regulation of cullin-RING ubiquitin ligases. Nat Rev Mol Cell Biol 6:9–20
137. Jones J, Wu K, Yang Y et al (2008) A targeted proteomic analysis of the ubiquitin-like modifier nedd8 and associated proteins. J Proteome Res 7:1274–1287
138. Yen HC, Elledge SJ (2008) Identification of SCF ubiquitin ligase substrates by global protein stability profiling. Science 322:923–929
139. Smith PG, Milhollen M, Traore T et al (2007) Antitumor activity of MLN4924, a novel small molecule inhibitor of Nedd8-activating enzyme, in pre-clinical models of lymphoma [abstract]. Proceedings of Molecular Targets and Cancer Therapeutics 2007;331
140. Milhollen M, Narayanan U, Amidon B et al (2008) MLN4924, a potent and novel small molecule inhibitor of Nedd8 activating enzyme, induces DNA re-replication and apoptosis in cultured human tumor cells [abstract]. Eur J Cancer 6:113
141. Berger AJ, Yu J, Garnsey J et al (2008) Pharmacodynamic and efficacy relationship of MLN4924, a novel small molecule inhibitor of Nedd8-activating enzyme, in human xenograft tumors grown in immunocompromised mice [abstract]. Eur J Cancer 6:29
142. Milhollen M, Narayanan U, Duffy J et al (2008) MLN4924, a potent and novel small molecule inhibitor of Nedd8 activating enzyme, induces DNA re-replication and apoptosis in cultured human tumor cells [abstract]. Blood 112: Abstract 3621

143. Milhollen M, Narayanan U, Berger AJ et al (2008) MLN4924, a novel small molecule inhibitor of Nedd8-activating enzyme, demonstrates potent anti-tumor activity in diffuse large B-cell lymphoma [abstract]. Blood 112: Abstract 606
144. Kupperman E, Lee EC, Cao Y et al (2010) Evaluation of the proteasome inhibitor MLN9708 in preclinical models of human cancer. Cancer Res 70:1970–1980
145. Zhou HJ, Aujay MA, Bennett MK et al (2009) Design and synthesis of an orally bioavailable and selective peptide epoxyketone proteasome inhibitor (PR-047). J Med Chem 52:3028–3038
146. Roccaro AM, Sacco A, Aujay M et al (2010) Selective inhibition of chymotrypsin-like activity of the immunoproteasome and constitutive proteasome in Waldenstrom macroglobulinemia. Blood 115:4051–4060
147. Muchamuel T, Basler M, Aujay MA et al (2009) A selective inhibitor of the immunoproteasome subunit LMP7 blocks cytokine production and attenuates progression of experimental arthritis. Nat Med 15:781–787
148. Kuhn DJ, Hunsucker SA, Chen Q et al (2009) Targeted inhibition of the immunoproteasome is a potent strategy against models of multiple myeloma that overcomes resistance to conventional drugs and nonspecific proteasome inhibitors. Blood 113:4667–4676
149. Rodler E, Infante JR, Shu L et al (2010) First-in-human, phase I dose-escalation study of investigational drug MLN9708, a second-generation proteasome inhibitor, in advanced nonhematologic malignancies [abstract]. J Clin Oncol 28(Suppl 15S):Abstract 3071

Index